Guide to Electricity and Magnetism

This is a "how to guide" for a calculus-based introductory course in electricity and magnetism. Students taking the subject at an intermediate or advanced level may also find it to be a useful reference. The calculations are performed in Mathematica, and stress graphical visualization, units, and numerical answers. The techniques show the student how to learn the physics without being hung up on the math. There is a continuing movement to introduce more advanced computational methods into lower-level physics courses. Mathematica is a unique tool in that code is written as "human readable" much like one writes a traditional equation on the board.

Key Features:

- Concise summary of the physics concepts.
- Over 300 worked examples in Mathematica.
- Tutorial to allow a beginner to produce fast results.

James Rohlf is a Professor at Boston University. As a graduate student he worked on the first experiment to trigger on hadron jets with a calorimeter, Fermilab E260. His thesis (G. C. Fox, advisor, C. Barnes, R. P. Feynman, R. Gomez) used the model of Field and Feynman to compare observed jets from hadron collisions to that from electron-positron collisions and made detailed acceptance corrections to arrive at first the measurement of quark-quark scattering cross sections. His thesis is published in Nuclear Physics B171 (1980) 1. At the Cornell Electron Storage Rings, he worked on the discovery of the Upsilon (4S) resonance and using novel event shape variables developed by Stephen Wolfram and his thesis advisor, Geoffrey Fox. He performed particle identification of kaons and charmed mesons to establish the quark decay sequence, b –> c. At CERN, he worked on the discovery of the W and Z bosons and measurement of their properties. Presently, he is working on the Compact Muon Solenoid (CMS) experiment at the CERN Large Hadron Collider (LHC) which discovered the Higgs boson and is searching for new phenomena beyond the standard model.

Guide to Electricity and Magnetism

Using Mathematica for Calculations and Visualizations

James W. Rohlf

CRC Press
Taylor & Francis Group
Boca Raton London New York

CRC Press is an imprint of the
Taylor & Francis Group, an **informa** business

Designed cover image: James W. Rohlf

First edition published 2024
by CRC Press
2385 NW Executive Center Drive, Suite 320, Boca Raton FL 33431

and by CRC Press
4 Park Square, Milton Park, Abingdon, Oxon, OX14 4RN

CRC Press is an imprint of Taylor & Francis Group, LLC

ISBN: 978-1-032-64667-1 (hbk)
ISBN: 978-1-032-64085-3 (pbk)
ISBN: 978-1-032-64668-8 (ebk)

Typeset in Nimbus font
by KnowledgeWorks Global Ltd.

DOI: 10.1201/9781032646688

Publisher's note: This book has been prepared from camera-ready copy provided by the authors.

Contents

Preface xiii

Chapter 1 ▪ Coulomb's Law and Electric Field 1

1.1 ELECTRIC CHARGE 1
1.2 INVERSE SQUARE LAW 2
1.3 VECTOR NATURE OF THE FORCE 4
1.3.1 Simple Coordinates 4
1.3.2 General Coordinates 5
1.4 SUPERPOSITION PRINCIPLE 6
1.4.1 Three Charges 6
1.4.2 Thirteen Charges Minus One 7
1.4.3 Finding the Location of Zero Force 8
1.5 ELECTRIC FIELD 12
1.5.1 Point Charge 12
1.5.2 General Formula 13
1.5.3 Line of Charge 13
1.5.4 Ring of Charge 17
1.5.5 Disk of Charge 19
1.5.6 Sphere of Charge 21
1.5.7 Ball of Charge 23
1.6 AVERAGE FIELD ON A SPHERE OR BALL 25
1.7 CURL OF THE FIELD 26
1.8 STRENGTH OF THE ELECTRIC FORCE 27

CHAPTER 2 ▪ Gauss's Law 29

2.1 ELECTRIC FLUX 29
2.2 FLUX THROUGH A CLOSED SURFACE 32
2.2.1 Charge at Center of a Sphere 32
2.2.2 Flux Through a Hemisphere 32
2.2.3 Flux Through a Cube 34
2.2.4 Flux Through a Sphere from a Charge Inside 35
2.2.5 Flux Through a Sphere from a Charge Outside 37
2.3 MAXWELL EQUATION 38
2.4 APPLYING GAUSS'S LAW 39
2.4.1 Line of Charge 40
2.4.2 Plane of Charge 41
2.4.3 Ball of Charge 42

CHAPTER 3 ▪ Electric Potential 45

3.1 ELECTRIC POTENTIAL DIFFERENCE 45
3.2 CURL OF E 45
3.3 POINT CHARGE 46
3.3.1 Field from the Potential 46
3.4 FORMULA FOR V 48
3.5 LINE OF CHARGE 48
3.6 RING OF CHARGE 50
3.7 DISK OF CHARGE 50
3.8 SPHERE OF CHARGE 51
3.9 BALL OF CHARGE 52
3.10 ELECTRIC DIPOLE 53
3.11 CONDUCTORS 56
3.11.1 General Properties 56
3.11.2 Field Near the Surface 56
3.12 CAPACITANCE 57
3.12.1 Capacitance of a Ball or Sphere 57
3.12.2 Parallel Plates 58
3.12.3 Coax cable 58

3.13 STORED ENERGY OF A CHARGE DISTRIBUTION 60
3.13.1 Point Charges 60
3.13.2 Stored Energy in a Capacitor 61
3.13.3 Stored Energy in Terms of the Field 62

CHAPTER 4 ▪ The Biot-Savart Law 63

4.1 MAGNETIC FORCE AND MAGNETIC FIELD 63
4.2 MAGNETIC FLUX 65
4.3 MAXWELL EQUATION 65
4.4 ELECTRIC CURRENT AND THE BIOT-SAVART LAW 66
4.5 CONTINUITY EQUATION 67
4.6 LINE OF CURRENT 68
4.7 CYLINDRICAL SHELL OF CURRENT 70
4.8 CURRENT LOOP 71
4.9 SOLENOID 73
4.10 MAGNETIC DIPOLE 75

CHAPTER 5 ▪ Ampere's Law 79

5.1 EXAMPLES OF AMPÈRE'S LAW 80
5.1.1 Square Loop 80
5.1.2 Circular Loop 82
5.2 DERIVATION FROM THE BIOT-SAVART LAW 83
5.3 APPLYING AMPÈRE'S LAW 85
5.3.1 Straight Line of Current 85
5.3.2 Inside a Long Cylinder of Current 86
5.3.3 Sheet of Current 87
5.3.4 Solenoid 87
5.3.5 Toroid 88
5.4 INCOMPLETE MAXWELL EQUATION 89

CHAPTER 6 ▪ Magnetic Vector Potential 91

6.1 DETERMINING **A** FROM **B** 92
6.1.1 Vector Potential of a Solenoid 92

6.1.2 Vector Potential of a Sheet of Current 93
6.1.3 Vector Potential of a Long Wire 94
6.2 CALCULATING **A** BY DIRECT INTEGRATION 94
6.2.1 Vector Potential of a Wire Segment 96
6.2.2 Vector Potential for a Spinning Sphere of Charge 97
6.2.3 Vector Potential for a Dipole 100

CHAPTER 7 ▪ Faraday's Law 103

7.1 THE FLUX RULE 103
7.2 MAXWELL EQUATION 104
7.3 MUTUAL INDUCTANCE 105
7.3.1 Two-Loop Example 106
7.3.2 Nested Solenoids 108
7.4 SELF-INDUCTANCE 109
7.5 STORED MAGNETIC ENERGY 110

CHAPTER 8 ▪ Circuits 113

8.1 OHM'S LAW 113
8.1.1 Electric Field 113
8.1.2 Drift Speed 114
8.2 CIRCUITS WITH RESISTORS 116
8.2.1 Resistors in Series 116
8.2.2 Resistors in Parallel 116
8.3 KIRCHHOFF'S RULES 117
8.3.1 Two-Loop Circuit 117
8.3.2 Three-Loop Circuit 118
8.3.3 $\Delta-Y$ Transform 120
8.4 RC CIRCUIT 121
8.5 ALTERNATING CURENT 122
8.6 DRIVEN *LR* CIRCUIT 122
8.7 DRIVEN *RC* CIRCUIT 125
8.8 DRIVEN *LRC* CIRCUIT 127

CHAPTER 9 ▪ Fields Inside Materials 131

9.1 POLARIZATION VECTOR 131
9.1.1 Bound Charge Density 132
9.1.2 Radially Polarized Cube 132
9.1.3 Uniformly Polarized Ball 134
9.2 DISPLACEMENT VECTOR D 136
9.2.1 Linear Dielectric 136
9.2.2 Dielectric Ball in an External Electric Field 137
9.3 CAPACITOR WITH A DIELECTRIC 138
9.4 POINT CHARGE NEAR A DIELECTRIC BOUNDARY 139
9.5 MAGNETIZATION VECTOR 140
9.5.1 Bound Currents 141
9.5.2 Uniformly Magnetized Ball 143
9.6 AUXILIARY FIELD H 146
9.6.1 Linear Materials 147
9.6.2 Magnetic Ball in an External Magnetic Field 148
9.7 BOUNDARY CONDITIONS 149

CHAPTER 10 ▪ Electromagnetic Waves 151

10.1 MODIFYING AMPÈRE'S LAW 151
10.2 MAXWELL'S EQUATIONS WITH A MAGNETIC CHARGE 152
10.3 POYNTING VECTOR 152
10.3.1 Current in a Wire 154
10.3.2 Charging Capacitor 155
10.4 THE WAVE EQUATION 156
10.5 PLANE WAVES 157
10.5.1 Time Average 159
10.5.2 Exponential Notation 159
10.6 WAVES IN MATTER 159
10.7 REFLECTION AND REFRACTION 161
10.8 OBLIQUE INCIDENCE 163

10.8.1 Relationships Between the Angles 164
10.8.2 Choosing the Polarization 165
10.8.3 Brewster Angle 167

CHAPTER 11 ▪ Fields in Moving Frames of Reference 171

11.1 BASICS OF SPECIAL RELATIVITY 171
11.2 FOUR-VECTORS 172
11.3 LORENTZ TRANSFORMATION 173
11.3.1 Time Dilation 174
11.3.2 Length Contraction 175
11.4 ENERGY-MOMENTUM FOUR-VECTOR 176
11.4.1 Speed 176
11.4.2 Total Energy 177
11.4.3 Kinetic Energy 178
11.4.4 Velocity Addition 179
11.5 EXAMPLES OF FIELDS IN MOVING FRAMES 180
11.5.1 Moving Capacitor 180
11.5.2 Steady Current 181
11.6 POINT CHARGE WITH CONSTANT VELOCITY 184
11.7 TRANSFORMATION OF THE FIELDS 186
11.8 MAGNETIC FIELD OF A MOVING POINT CHARGE 189

APPENDIX A ▪ Mathematica Starter 191

A.1 CELLS 191
A.2 PALETTES 192
A.3 FUNCTIONS 192
A.4 RESERVED NAMES 193
A.5 PHYSICAL CONSTANTS AND THEIR UNITS 194

APPENDIX B ▪ Vectors 197

B.1 NOTATION 197
B.1.1 Scalar Product 197
B.1.2 Vector Product 198

B.2 DERIVATIVES 198
B.2.1 Gradient 198
B.2.2 Divergence 199
B.2.3 Curl 199
B.3 LINE INTEGRAL 200
B.4 FLUX INTEGRAL 202
B.5 DIVERGENCE THEOREM 203
B.6 STOKES' THEOREM 204

APPENDIX C ▪ Spherical and Cylindrical Coordinates 205

C.1 SPHERICAL COORDINATES 205
C.2 CYLINDRICAL COORDINATES 208

APPENDIX D ▪ Physical Constants 211

Index 215

Preface

This guide is designed to be used by students in a calculus-based introductory course in electricity and magnetism. Students taking the subject at an intermediate or advanced level may also find it to be a useful reference.

Appendix A offers a help to begin using Mathematica with no prior knowledge. Jumping right in and typing things, making your own mistakes, and then making use of the extensive inline documentation is a great way to learn. The beginner will want to keep a notebook (.nb file) to cut and paste from to avoid retyping of recurring expressions.

The first calculation in electric phenomena is likely to be using Coulomb's law, which needs for its input the fundamental electric charge and the electric force constant as well as a distance scale. Even a simple calculation is not likely to be done without a calculator. Example 1.6 appears in Chap. 1.

Example 1.6 Calculate the magnitude of the force between 2 protons separated by a distance of 1 nm.

```
In[6]:= r = 1 nm; N[UnitConvert[e^2/(4 π ε0 r^2), N], 3]

Out[6]= 2.31×10^-10 N
```

The units have been specified in newtons (N) and if the expression does not match units, it will not execute. This alone is worth its weight in gold for both the beginner and the expert alike. Mathematica is the ultimate physics calculator. The code can be used as a template for additional calculations and is available for download.

A second place that Mathematica excels is in the calculation of line and surface integrals that are needed for the concepts of electric potential and flux. In an example from Chap 2, a charge is placed at an arbitrary position inside a sphere.

Example 2.5 Calculate the electric flux through the sphere when the charge is inside.

```
In[8]:= $Assumptions = {0 < z < a}; r′ = {0, 0, z};
        r = a {Sin[θ] Cos[ϕ], Sin[θ] Sin[ϕ], Cos[θ]};
        ℛ = r - r′;
        f[θ_] =
          Integrate[Integrate[(q ℛ / (4 π ε0 (ℛ.ℛ)^(3/2))).(r/a) a^2 Sin[θ],
            {ϕ, 0, 2 π}], θ];
        Simplify[f[π] - f[0]]

Out[10]= q/ε0
```

Magnetic phenomena are famously confusing for beginners because there are so many vectors involved due to the motion of charge. Examples are carefully chose such that they can be used as templates to evaluate integrals containing non-trivial operations such as multiple vector products.

Finally, Mathematica is extremely useful for its algebraic manipulation of complicated expressions and to solve a system of simultaneous equations such as appears in circuits with multiple branch points.

CHAPTER 1

Coulomb's Law and Electric Field

Three fundamental physical constants are encountered in electricity and magnetism, the elementary charge (e) of Sect. 1.1, the electric constant (ε_0) of Sect. 1.2, and the magnetic constant (μ_0) of Chap. 4. They are given to four significant figures in Tab. 1.1.

Table 1.1 Fundamental constants of electricity and magnetism.

Constant	Value
Elementary charge (e)	1.602×10^{-19} C
Electric constant (ε_0)	8.854×10^{-12} C/(m V)
Magnetic constant (μ_0)	1.257×10^{-6} m T/A

1.1 ELECTRIC CHARGE

The source of the electric field is the electric charge. Charge is an intrinsic property of particles. Electrons have negative charge and protons have positive charge. The symbol e is used to represent the proton charge, also referred to as the "elementary charge." It is the first fundamental constant listed in Tab. 1.1. The electron charge is $-e$. The symbols q and Q are often used to represent electric charge of arbitrary sign and magnitude. The elementary charge is acquired with the function Quantity["ElementaryCharge"].

Example 1.1 Get the elementary charge.

```
In[1]:= Quantity["ElementaryCharge"]

Out[1]= e
```

The output of Ex. 1.1 is written in a slightly different shade in Mathematicata to distinguish its use as a physical quantity with units. The unit of charge is the coulomb (C). The coulomb unit is acquired with Quantity["Coulombs"].

Example 1.2 Get the coulomb unit.

```
In[2]:= Quantity["Coulombs"]

Out[2]= 1 C
```

The function UnitConvert[*quantity*, *targetunit*] converts the expression in the first argument to the unit specified in the second argument. It is a very powerful feature of Mathematica that the code will fail if the unit of the expression does not match the specified unit. The default unit is in the International System (SI). In the strict definition of SI, the unit C is derived from amp (A) second (s), so C must be specified or else the output will be in A · s. The function N[*expr*,*n*] gives a numerical value of the expression to *n* figures.

Example 1.3 Get the numerical value of the elementary charge in C to 10 figures.

```
In[3]:= N[UnitConvert[e, C], 10]

Out[3]= 1.602176634 × 10^-19 C
```

1.2 INVERSE SQUARE LAW

The experimental result known as Coulomb's law says that the electric force between two charges q_1 and q_2 is proportional to each of the charges and inversely proportional to the square of the separation distance r. The magnitude of the force F is written

$$F = \frac{kq_1q_2}{r^2},$$

where k is a constant of proportionality to be determined from measurement. Historically, the constant is written

$$k = \frac{1}{4\pi\varepsilon_0},$$

where ε_0 is called the electric constant. It is the second fundamental constant listed in Tab. 1.1. The electric constant is obtained with Quantity["ElectricConstant"].

Example 1.4 Get the electric constant.

```
In[4]:= Quantity["ElectricConstant"]

Out[4]= ε0
```

Example 1.5 Get the numerical value of $\frac{1}{4\pi\varepsilon_0}$.

```
In[5]:= N[UnitConvert[1/(4 π ε0), (N m^2)/C^2], 3]

Out[5]= 8.99×10^9 m^2N/C^2
```

With the definition of electric potential in Chap. 3 and the concept of electric field (Sect. 1.5), a useful unit for the electric constant is $\frac{\mathrm{C}}{\mathrm{m}\cdot\mathrm{V}}$ which is equal to $\frac{\mathrm{N}\cdot\mathrm{m}^2}{\mathrm{C}^2}$.

A semicolon after a command suppresses the output. In Ex. 1.6, the variable r is set equal to 1 nm but this result is not output.

Example 1.6 Calculate the magnitude of the force between two protons separated by a distance of 1 nm.

```
In[6]:= r = 1 nm; N[UnitConvert[e^2/(4 π ε0 r^2), N], 3]

Out[6]= 2.31×10^-10 N
```

1.3 VECTOR NATURE OF THE FORCE

1.3.1 Simple Coordinates

Consider charges q_1 and q_2 separated by a distance $\mathbf{r}$ as shown in Fig. 1.1. Take the origin to be at the location of q_1. The force on q_2 caused by q_1 is

$$\mathbf{F} = \frac{q_1 q_2}{4\pi\varepsilon_0 r^2}\hat{\mathbf{r}},$$

where $\hat{\mathbf{r}}$ is the unit vector in the direction of $\mathbf{r}$ which points from q_1 to q_2. Note that the algebra automatically takes care of the sign of the charges. If the charges are the same sign, the force is repulsive (along $\hat{\mathbf{r}}$). If the charges have the opposite sign, the force is attractive (along $-\hat{\mathbf{r}}$). For this to hold true, all one needs to remember is that the vector r points from the "source" to the place where the force (or field of Sect. 1.5) is calculated. The charge q_1 is the source of the force on q_2.

Newton's third law pair is the force on q_1 caused by q_2. This force is obtained by placing the origin at q_2 as it becomes the source of the force on q_1.

Using $\hat{\mathbf{r}} = \mathbf{r}/r$, it is useful to write the force as

$$\mathbf{F} = \frac{q_1 q_2}{4\pi\varepsilon_0 r^3}\mathbf{r}.$$

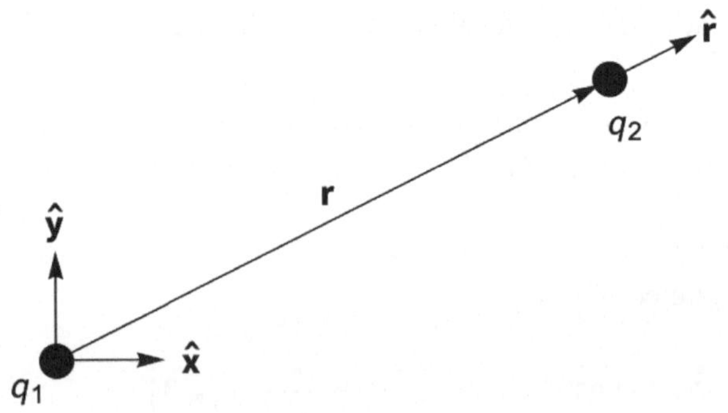

Figure 1.1 Diagram for the Coulomb force between two charges. The origin is placed at the location of q_1. The vector $\mathbf{r}$ points to q_2. The force on q_2 caused by q_1 is in the direction $\hat{\mathbf{r}}$ ($-\hat{\mathbf{r}}$) if the charges are the same (opposite) sign.

1.3.2 General Coordinates

The physics cannot depend on the choice of coordinates. It is extremely useful and often essential to have an arbitrary choice of origin as shown in Fig. 1.2. The vector $\mathbf{r}'$ points to the source charge q_1, and the vector $\mathbf{r}$ points to q_2. The vector that points from q_1 to q_2 is

$$\boldsymbol{\mathcal{R}} = \mathbf{r} - \mathbf{r}'.$$

Coulomb's law reads

$$\mathbf{F} = \frac{q_1 q_2}{4\pi\varepsilon_0 |\mathbf{r} - \mathbf{r}'|^3}(\mathbf{r} - \mathbf{r}') = \frac{q_1 q_2}{4\pi\varepsilon_0 \mathcal{R}^2}\hat{\boldsymbol{\mathcal{R}}} = \frac{q_1 q_2}{4\pi\varepsilon_0 \mathcal{R}^3}\boldsymbol{\mathcal{R}}.$$

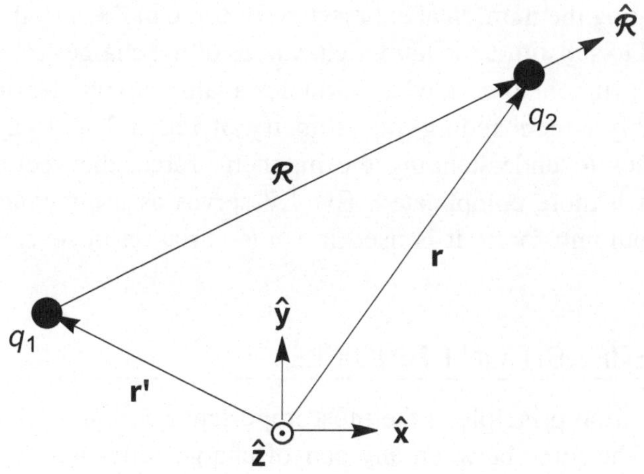

Figure 1.2 Diagram for the Coulomb force between 2 charges. The origin is placed at an arbitrary location. The vector $\mathbf{r}'$ points to q_1 and the vector $\mathbf{r}$ points to q_2. The force on q_2 caused by q_1 is in the direction $\hat{\boldsymbol{\mathcal{R}}}$ ($-\hat{\boldsymbol{\mathcal{R}}}$) if the charges are the same (opposite) sign.

The 3 vectors of Fig. 1.2 form a triangle with $\mathbf{r}' + \boldsymbol{\mathcal{R}} = \mathbf{r}$. A vector squared is the square root of the dot product of the vector with itself. The magnitude of $\boldsymbol{\mathcal{R}}$ is

$$\mathcal{R} = \sqrt{(\mathbf{r} - \mathbf{r}')^2} = \sqrt{r^2 + r'^2 - 2\mathbf{r} \cdot \mathbf{r}'} = \sqrt{r^2 + r'^2 - 2rr' \cos\theta},$$

where θ is the angle between $\mathbf{r}$ and $\mathbf{r}'$. This important result is known as the law of cosines.

A vector is represented in Mathematica with $\{x, y, z\}$ (see App. B). The dot product of vectors, in this case $\boldsymbol{\mathcal{R}} \cdot \boldsymbol{\mathcal{R}}$, is written $\mathcal{R}.\mathcal{R}$, and $\mathcal{R}^3$ is written $(\mathcal{R}.\mathcal{R})^{3/2}$.

Example 1.7 Charge $q_1 = 1$ nC is located at $(x,y,z) = (1, 0, 0)$ m and $q_2 = 3$ nC is at $(1, 2, 3)$ m. Calculate the force vector that q_1 exerts on q_2.

```
In[7]:= r′ = {1, 0, 0} m; r = {1, 2, 3} m; ℛ = r - r′;
        F =
         N[UnitConvert[ q1 q2 ℛ / (4 π ε0 (ℛ.ℛ)^(3/2)) /. {q1 → 1 nC, q2 → 3 nC},
           nN], 3]
Out[8]= {0 nN, 1.15 nN, 1.73 nN}
```

In evaluating the numerical expression for force in Ex. 1.7, the command "/." was used to substitute the numerical values of the charges q_1 and q_2. This allows those symbols to be used as variables again without clearing them.

Thoroughly understanding the geometry of Fig. 1.2 will pay dividends when it comes to understanding the magnetic force, the vector nature of which is much more complicated. Ex. 1.7 serves as a template for calculating the Coulomb force. It is used in some variation in several following examples.

1.4 SUPERPOSITION PRINCIPLE

The superposition principle is the most important concept in electricity and magnetism. The force between any pair of charges does not depend on the presence of other charges.

1.4.1 Three Charges

As mentioned, the code of Fig. 1.7 can be used as a template for more complicated calculations. Figure 1.3 shows two source charges for the force. One can make position vectors for each of them, $\mathbf{r}_1'$ and $\mathbf{r}_2'$, calculate $\mathcal{R}_1$ and $\mathcal{R}_2$, and then add together the contributions to the total force.

Example 1.8 Calculate the force on q_2 from the 2 q_1 charges in the geometry of Fig. 1.3.

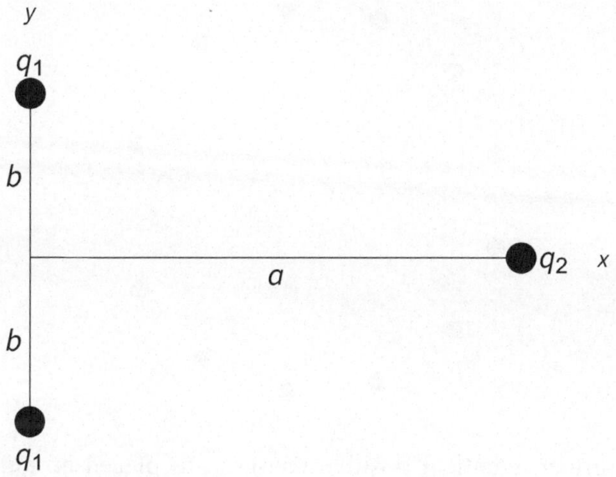

Figure 1.3 Each charge q_1 acts on q_2 independently. The net charge of q_2 is calculated in Ex. 1.8.

```
In[9]:= r′1 = {0, b, 0}; r′2 = {0, -b, 0};
        r = {a, 0, 0}; ℛ1 = r - r′1; ℛ2 = r - r′2;
```

$$F = \frac{q_1\, q_2\, \mathcal{R}_1}{4\,\pi\,\varepsilon_0\,(\mathcal{R}_1 . \mathcal{R}_1)^{3/2}} + \frac{q_1\, q_2\, \mathcal{R}_2}{4\,\pi\,\varepsilon_0\,(\mathcal{R}_2 . \mathcal{R}_2)^{3/2}}$$

Out[11]= $\left\{\frac{a\, q_1\, q_2}{2\,(a^2+b^2)^{3/2}\,\pi\,\varepsilon_0},\ 0,\ 0\right\}$

Example 1.8 demonstrates a simple case of symbolic manipulation at which Mathematica excels without peer.

1.4.2 Thirteen Charges Minus One

Thirteen identical positive charges q are placed at equal intervals around the circumference of a circle of radius R. From symmetry, the electric field at the center of the circle is zero. Now remove one of the charges (Fig. 1.4). The potential due to the 12 charges by superposition must be equal to that due to the 13 positive charges plus a negative charge $-q$ placed at the location of the charge that was removed.

Example 1.9 Calculate the electric field at the center of the circle from all 13 charges and also with one of the charges removed as indicated in Fig. 1.4.

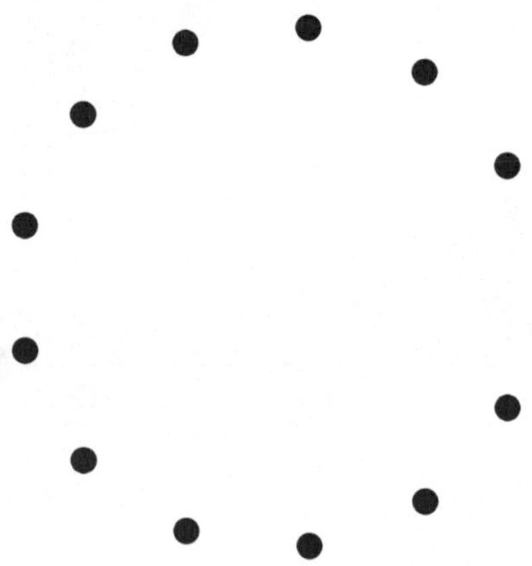

Figure 1.4 Thirteen identical positive charges are placed at equal intervals around the circumference of a circle, and then one of them is removed.

```
In[12]:= θ = 2 π / 13;
         r = Table[-{Cos[n θ], Sin[n θ], 0}, {n, 13}];
```

$$\sum_{j}^{13} \frac{q\, r[[j]]}{4 \pi \varepsilon_0 R^2} \text{ // FullSimplify}$$

$$\sum_{j}^{12} \frac{q\, r[[j]]}{4 \pi \varepsilon_0 R^2} \text{ // FullSimplify}$$

```
Out[13]= {0, 0, 0}
```

$$\text{Out[14]=} \left\{\frac{q}{4 \pi R^2 \varepsilon_0}, 0, 0\right\}$$

The function Table[ ] provides a convenient way to do a repetitive calculation.

The electric field from all 13 charges is indeed zero. The electric field with the charge on the x-axis removed is in the x direction and is positive for positive q, pointing at the missing charge, at the position where a negative charge would be placed together with the original 13 charges to solve the problem by superposition.

1.4.3 Finding the Location of Zero Force

In another type of superposition problem, suppose two charges $2q$ and $-q$ are separated by a distance $2a$ as shown in Fig. 1.5. Is there a place where a third charge would feel zero force? If so, that location must be on the x-axis.

The function Solve[*expr, var, domain*] will solve the expression in the first argument, for the variable in the second argument, over the domain specified in the third argument.

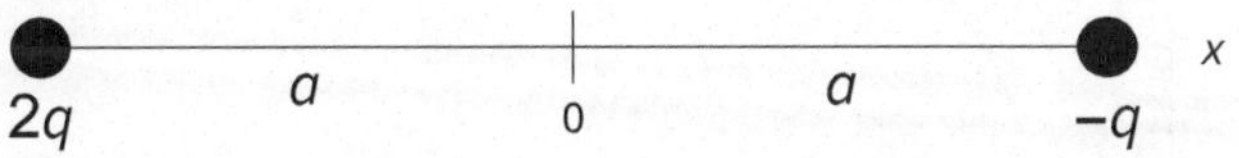

Figure 1.5 Charge $2q$ is located at $x = -a$, and $-q$ is located at $x = a$. The location where the force on a third charge is zero is solved in Ex. 1.10.

Example 1.10 Find the location where the force on a 3rd charge is zero. Let the third charge be Q at $x = d$.

```
In[15]:= << Notation`
         Symbolize[ParsedBoxWrapper[SubscriptBox["_", "_"]]]
         $Assumptions = a > 0; r'₁ = {-a, 0, 0}; r'₂ = {a, 0, 0};
         r = {d, 0, 0};
         ℛ₁ = r - r'₁; ℛ₂ = r - r'₂; q₁ = 2 q; q₂ = -q;
         F = q₁ Q ℛ₁ / (4 π ε₀ (ℛ₁.ℛ₁)^(3/2)) + q₂ Q ℛ₂ / (4 π ε₀ (ℛ₂.ℛ₂)^(3/2));
         Solve[F == 0, d, Reals]
Out[19]= {{d → 3 a + 2 √2 a}}
```

Note that in Ex. 1.10 one did not have to specify if d is positive or negative. The solution determines that. The parameter a is specified to be greater than zero with the $Assumptions command. One has to be careful with the use of subscripts in Mathematica code such that they do not interfere with components of vectors. The first two lines of code are protecting the use of subscripts as serial numbers such that they do not interfere with the vector operations in Solve. They produce no output.

Figure 1.6 shows the force on a third charge placed to the right of $-q$ (see Fig. 1.5). The force is zero at a bit less than $6a$ as calculated exactly in Ex. 1.10. At large distnces, the force looks like that of a single charge q at the origin.

Figure 1.7 illustrates a more involved calculation that is nevertheless straightforward in Mathematica. Three equal charges are placed at the vertices of an equilateral triangle. Find the place(s) where a fourth charge would experience zero force. Clearly this cannot occur outside the triangle where all charges would contribute to either repulsion or attraction.

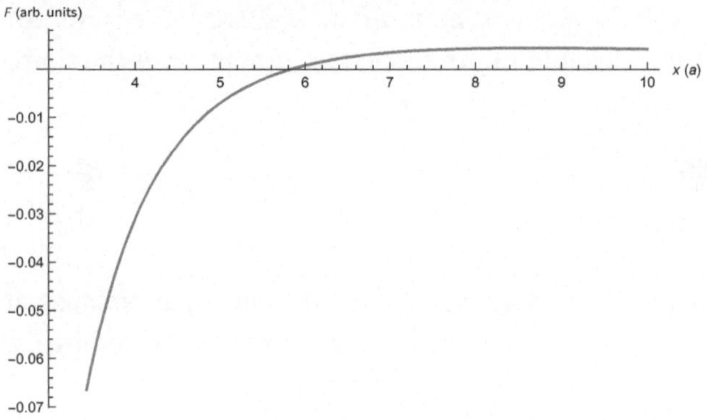

Figure 1.6 Force on a third charge (see Fig. 1.5) as a function of distance along the x-axis from the origin.

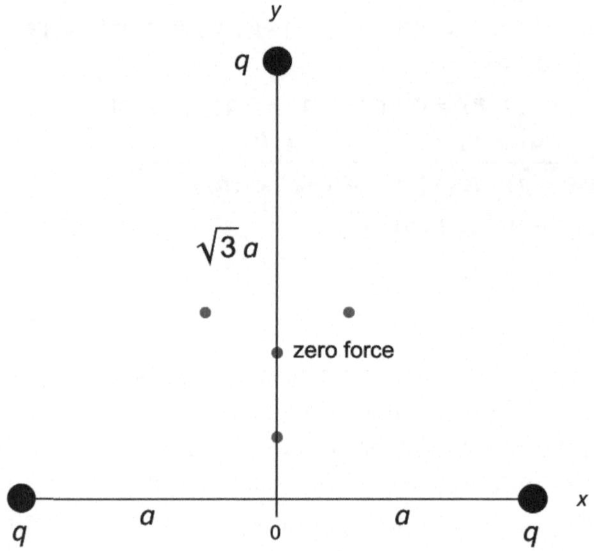

Figure 1.7 Charge q is located at $(x,y) = (-a,0)$, $(a,0)$, and $(0, \sqrt{3}\,a)$ forming an equilateral triangle. There are two locations along the vertical axis where the force on a fourth charge would be zero as solved in Ex. 1.11 and two more from symmetry.

Example 1.11 Find the solution along the y-axis of Fig. 1.7 where the force would be zero on a fourth charge. Let d be this distance. Get the numerical value.

```
In[20]:= $Assumptions = {a > 0}; r′1 = {-a, 0, 0};
         r′2 = {a, 0, 0};  r′3 = {0, √3 a, 0};
         r = {0, d, 0};
         ℛ1 = r - r′1; ℛ2 = r - r′2; ℛ3 = r - r′3;
         F = q Q ℛ1/(4 π ε0 (ℛ1.ℛ1)^(3/2)) + q Q ℛ2/(4 π ε0 (ℛ2.ℛ2)^(3/2)) + q Q ℛ3/(4 π ε0 (ℛ3.ℛ3)^(3/2));
         f = Solve[F == 0, d, Reals] // FullSimplify
         N[f /. a → 1]
```

```
Out[22]= {{d → Root[-3 a^8 + 2 √3 a^7 #1 + 98 a^6 #1^2 -
            210 √3 a^5 #1^3 + 528 a^4 #1^4 - 234 √3 a^3 #1^5 +
            174 a^2 #1^6 - 22 √3 a #1^7 + 3 #1^8 &, 2]}, {d → a/√3}}

Out[23]= {{d → 0.248586}, {d → 0.57735}}
```

Two solutions are found for d. The first one cannot be put in closed form. The second one is what was expected from symmetry.

Example 1.12 Find the place on the y-axis of Fig. 1.7 where the force is a local maximum (other than the singularity at $y = \sqrt{3}\,a$). Output the force, its derivative, and the numerical solution. The expression F[[n]] is taking the nth component of the vector **F**.

```
In[20]:= F[[2]]
         D[F[[2]] /. a → 1, d]
         f = Solve[D[F[[2]] /. a → 1, d] == 0, d, Reals]; N[f /. a → 1]
```

$$\text{Out[20]=}\ \frac{\left(-\sqrt{3}\,a+d\right) q\,Q}{4\left(\left(-\sqrt{3}\,a+d\right)^2\right)^{3/2}\pi\,\varepsilon_0}+\frac{d\,q\,Q}{2\left(a^2+d^2\right)^{3/2}\pi\,\varepsilon_0}$$

$$\text{Out[21]=}\ -\frac{3\left(-\sqrt{3}+d\right)^2 q\,Q}{4\left(\left(-\sqrt{3}+d\right)^2\right)^{5/2}\pi\,\varepsilon_0}+\frac{q\,Q}{4\left(\left(-\sqrt{3}+d\right)^2\right)^{3/2}\pi\,\varepsilon_0}-\frac{3\,d^2\,q\,Q}{2\left(1+d^2\right)^{5/2}\pi\,\varepsilon_0}+\frac{q\,Q}{2\left(1+d^2\right)^{3/2}\pi\,\varepsilon_0}$$

```
Out[22]= {{d → -0.641476}, {d → 0.416009}}
```

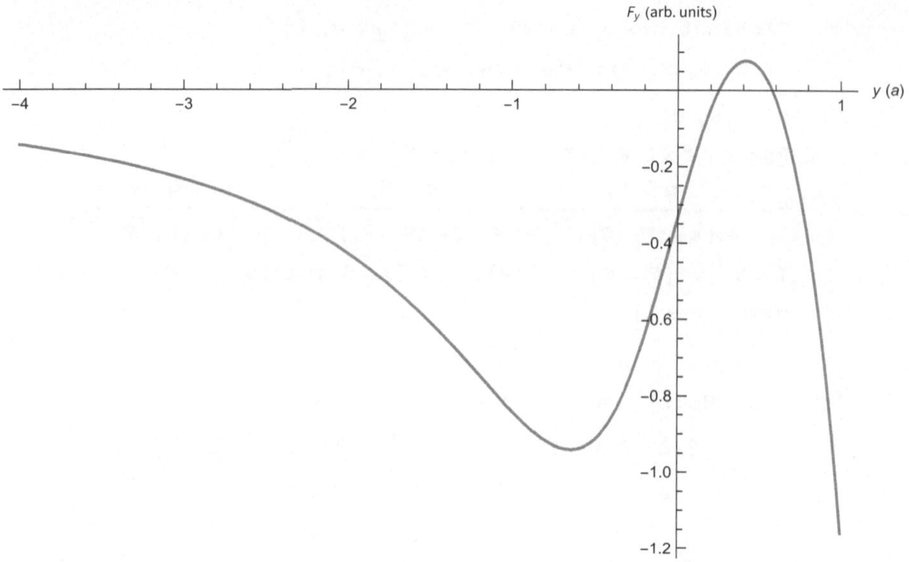

Figure 1.8 The plot of the force on a fourth charge *vs.* y of Fig. 1.7 crosses the horizontal-axis twice giving the two locations where the force is zero as solved in Ex. 1.11. There are two more locations of zero force from symmetry as indicated in Fig. 1.7.

The negative solution is the local minimum of Fig. 1.8, and the positive solution is the local maximum that occurs between the zeros.

1.5 ELECTRIC FIELD

The concept of electric field is born from the superposition principle.

1.5.1 Point Charge

In the example of the Coulomb force between two charges (Fig. 1.1), one can factor out q_2 to get

$$\mathbf{F} = q_2\left(\frac{q_1}{4\pi\varepsilon_0 r^2}\hat{\mathbf{r}}\right) = q_2\mathbf{E},$$

where

$$\mathbf{E} = \frac{q_1}{4\pi\varepsilon_0 r^2}\hat{\mathbf{r}}$$

is the electric field due to q_1. For any distribution of charges, the force that an additional "test" charge would get is $q\mathbf{E}$ where $\mathbf{E}$ is the electric field. Note that the test charge is not part of the definition of electric field. The unit of the electric field is N/C.

In Ex. 1.10 and Sect. 1.11, one determined the places where the electric field was zero. The electric field has a deep and fundamental role in electricity and magnetism as an ingredient of the electromagnetic wave (Chap. 10). The above expression for the electric field of a point charge has the coordinate system at the location of the point charge. If there are multiple charges, one can no longer do this, and the notation of Fig. 1.2 is useful.

1.5.2 General Formula

From superposition, we can write the electric field for a line of charge with density λ as

$$\mathbf{E} = \frac{1}{4\pi\varepsilon_0}\int d\ell' \frac{\lambda}{\mathcal{R}^2}\hat{\boldsymbol{\mathcal{R}}},$$

for a surface charge with density σ as

$$\mathbf{E} = \frac{1}{4\pi\varepsilon_0}\int da' \frac{\sigma}{\mathcal{R}^2}\hat{\boldsymbol{\mathcal{R}}},$$

and for a volume charge with density ρ as

$$\mathbf{E} = \frac{1}{4\pi\varepsilon_0}\int dv' \frac{\rho}{\mathcal{R}^2}\hat{\boldsymbol{\mathcal{R}}}.$$

The primed vector points to the location of the charge and is the integration variable and the vector $\boldsymbol{\mathcal{R}}$ is defined by Fig. 1.1 with the charge q_2 replaced by point $\mathcal{P}$, the place where the field is being calculated.

1.5.3 Line of Charge

Consider a line of charge (Fig. 1.9) that stretches along the x-axis from $-L$ to L. The charge per length is λ.

The electric field is calculated by dividing the charge into infinitesimal pieces dq,

$$dq = \lambda dx'.$$

At $x = 0$ and a distance y from the line of charge, the infinitesimal piece of the field is

$$d\mathbf{E} = \frac{dq}{4\pi\varepsilon_0 r^2}\hat{\mathbf{r}} = \frac{\lambda dx'}{4\pi\varepsilon_0(x'^2+y^2)}\frac{x'}{\sqrt{x'^2+y^2}}\hat{\mathbf{x}} + \frac{\lambda dx'}{4\pi\varepsilon_0(x'^2+y^2)}\frac{y}{\sqrt{x'^2+y^2}}\hat{\mathbf{y}}.$$

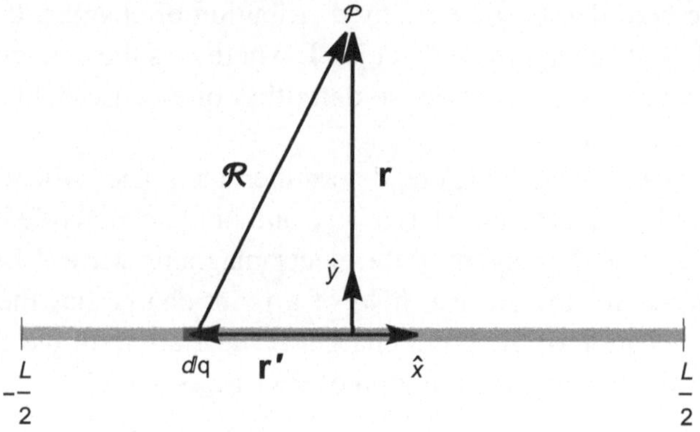

Figure 1.9 For a line of charge, the vector $\mathbf{r}'$ pointing to the charge becomes the integration variable. Perpendicular to the midpoint, the x coordinate of the observation point $\mathcal{P}$ is 0.

Notice that x' is the integration variable, y is the coordinate of point $\mathcal{P}$, and they are both needed to specify the distance between point $\mathcal{P}$ and a differential piece of the charge. The total field is

$$\mathbf{E} = \int_{-L/2}^{L/2} \left[\frac{\lambda dx'}{4\pi\varepsilon_0(x'^2+y^2)} \frac{x'}{\sqrt{x'^2+y^2}}\hat{\mathbf{x}} + \frac{\lambda dx'}{4\pi\varepsilon_0(x'^2+y^2)} \frac{y}{\sqrt{x'^2+y^2}}\hat{\mathbf{y}} \right].$$

This is an easy calculation with Mathematica because the vector integration (all three components) is done in one step as long as the vector $\mathcal{R}$ that points to $\mathcal{P}$ is properly defined.

Example 1.13 Calculate the field perpendicular to the midpoint of a line of charge. The distance is specified to be in the real domain ($y \in \mathbb{R}$) and not equal to 0.

```
In[23]:= $Assumptions = {L > 0, y ∈ ℝ, y ≠ 0};
         r = {0, y, 0}; r′ = {x′, 0, 0}; ℛ = r - r′;
```

$$\mathbf{E} = \frac{\lambda}{4\pi\varepsilon_0} \int_{-L/2}^{L/2} \frac{\mathcal{R}}{(\mathcal{R}.\mathcal{R})^{3/2}} \, dx'$$

$$\text{Out[25]=} \left\{0, \frac{L\lambda}{2\pi y\sqrt{L^2+4y^2}\,\varepsilon_0}, 0\right\}$$

Note that the Roman E is reserved for the exponential function *e*. The Greek Epsilon is used here which looks similar in Mathematica. One should be a little careful with this. The layout of the integral is taken from the "Basic Math Assistant" palette. Alternately, one may use the function Integrate[*f*, *z*].

At large values of *y*, the field is that of a point charge, $Q = \lambda L$. The function Series[$f, \{x, x_0, n\}$] generates a power series expansion of the function f about $x = x_0$ to order n.

Example 1.14 Calculate the field for the line charge of Fig. 1.9 at large values of *y*.

```
In[26]:= Series[Ε, {y, ∞, 3}] /. λ → Q/L
```

$$\text{Out[26]=} \left\{0, \frac{Q}{4\pi\varepsilon_0 y^2} + O\left[\frac{1}{y}\right]^4, 0\right\}$$

To get the answer for a very long line of charge, it is easy to take the limit as $L \to \infty$.

Example 1.15 Calculate the limit as $L \to \infty$.

```
In[27]:= Limit[Ε, L → ∞]
```

$$\text{Out[27]=} \left\{0, \frac{\lambda}{2\pi y \varepsilon_0}, 0\right\}$$

If the observation point is not on line perpendicular to the midpoint (Fig. 1.10), the integration to get the electric field is non-trivial. The change in code needed for the Mathematica integration is just one letter in order to change the observation point from $(0, y, 0)$ to $(x, y, 0)$.

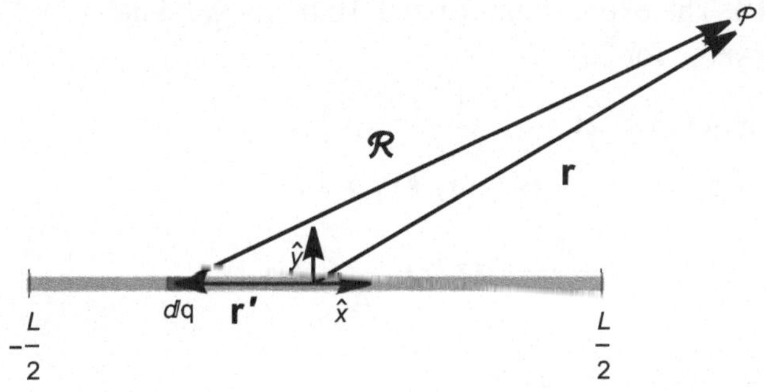

Figure 1.10 The *x* and *y* coordinates of the observation point $\mathcal{P}$ are arbitrary.

Example 1.16 Calculate the field off-axis due to a line of charge.

```
In[28]:= $Assumptions = {L > 0, y ∈ ℝ, y ≠ 0, x ∈ ℝ};
         r = {x, y, 0}; r' = {x', 0, 0}; ℛ = r - r';
```

$$\mathbb{E} = \frac{\lambda}{4\pi\varepsilon_0}\int_{-L/2}^{L/2}\frac{\mathcal{R}}{(\mathcal{R}.\mathcal{R})^{3/2}}\,dx' \text{ // Together // FullSimplify}$$

$$\text{Out[30]= }\left\{\frac{\left(-\sqrt{(L-2x)^2+4y^2}+\sqrt{(L+2x)^2+4y^2}\right)\lambda}{2\pi\sqrt{L^4+8L^2(-x^2+y^2)+16(x^2+y^2)^2}\,\varepsilon_0},\right.$$

$$\left(\left(2x\left(\sqrt{(L-2x)^2+4y^2}-\sqrt{(L+2x)^2+4y^2}\right)+L\left(\sqrt{(L-2x)^2+4y^2}+\sqrt{(L+2x)^2+4y^2}\right)\right)\lambda\right)\Big/$$

$$\left.\left(4\pi y\sqrt{L^4+8L^2(-x^2+y^2)+16(x^2+y^2)^2}\,\varepsilon_0\right),\ 0\right\}$$

The command Together[*expr*] makes a common denominator and FullSimplify[*expr*] returns the simplest form it can find. The // is shorthand to operate on the previous expression. This calculation will not be found in any ordinary textbook. It is too hard to do without computerized symbolic manipulation. The electric field vector is shown in Fig. 1.11.

Example 1.17 Take the limit of the field calculated in Ex. 1.16 to reproduce the result of Ex. 1.13.

```
In[25]:= Limit[E, x → 0] // Simplify
```

$$\text{Out[25]= }\left\{0,\ \frac{L\lambda}{2\pi y\sqrt{L^2+4y^2}\,\varepsilon_0},\ 0\right\}$$

Example 1.18 The expression of Ex. 1.16 in not valid at $y = 0$. Calculate the field on the x-axis. for $|x| > L/2$.

$$\text{In[31]:= \$Assumptions} = \left\{L > 0,\ -\frac{L}{2} < x > \frac{L}{2}\right\};$$

```
         r = {x, 0, 0}; r' = {x', 0, 0}; ℛ = r - r';
```

$$\mathbb{E} = \frac{\lambda}{4\pi\varepsilon_0}\int_{-L/2}^{L/2}\frac{\mathcal{R}}{(\mathcal{R}.\mathcal{R})^{3/2}}\,dx'$$

$$\text{Out[33]= }\left\{-\frac{L\lambda}{\pi(L^2-4x^2)\,\varepsilon_0},\ 0,\ 0\right\}$$

The function Simplify[*expr*] runs faster than FullSimplify[*expr*] and does the same thing for many cases.

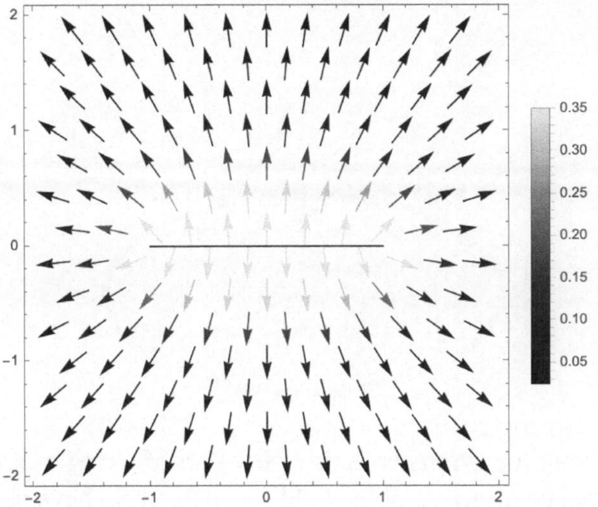

Figure 1.11 The vector electric field from a finite line of charge.

1.5.4 Ring of Charge

For a ring of charge (Fig. 1.12), along the symmetry axis, the electric field is straightforward because every part of the charge is at the same distance, and by symmetry the field points in the z direction. If the radius of the ring is a and the ring lies in the $x-y$ plane, then

$$\mathbf{r}' = a\cos\phi'\hat{\mathbf{x}} + a\sin\phi'\hat{\mathbf{y}},$$

where the angle ϕ' runs from 0 to 2π.

Example 1.19 Calculate the electric field on the symmetry axis for a ring of charge.

```
In[35]:= $Assumptions = {a > 0, z ∈ ℝ};
         r = {0, 0, z}; r' = {a Cos[ϕ'], a Sin[ϕ'], 0}; ℛ = r - r';
```

$$\mathbb{E} = \frac{\lambda a}{4\pi\varepsilon_0}\int_0^{2\pi}\frac{\mathcal{R}}{(\mathcal{R}.\mathcal{R})^{3/2}}\,d\phi'$$

$$\text{Out[37]=}\ \left\{0,\ 0,\ \frac{a z \lambda}{2\left(a^2+z^2\right)^{3/2}\varepsilon_0}\right\}$$

In the limit where $z \to \infty$, the result must give that of a point charge $Q = 2\pi a\lambda$.

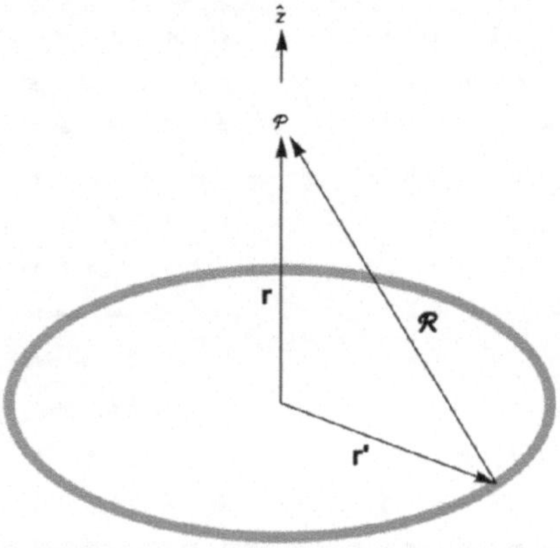

Figure 1.12 Along the symmetry axis, every part of a ring of charge is at the same distance. The direction of the field for each piece, however, is different.

Example 1.20 Calculate the limit a $z \to \infty$ for a ring of charge (on axis).

```
In[38]:= Series[E, {z, ∞, 2}] /. λ → Q/(2 π a)

Out[38]= {0, 0, Q/(4 π ε0 z^2) + O[1/z]^3}
```

The field off axis is a much harder problem. There is no closed-form solution. The easiest way to look at the solution is to expand in powers of $r = \sqrt{x^2+y^2}$. In Mathematica this can be done with AsymptoticIntegrate[f,{x,a,b},{x,x_0,n}] which calculates the definite integral expanding about x_0 to order n.

Example 1.21 Get the field for a ring of charge (off axis) for large r. There is ϕ symmetry so take $y = 0$ with out loss of generality.

```
In[39]:=
    Clear[r];
    $Assumptions = {R ∈ Reals, R > 0, r ∈ Reals, r > 0, φ ∈ Reals, φ > 0};
    ℛ = r {Sin[θ], 0, Cos[θ]} - a {Cos[φ′], Sin[φ′], 0};
    E = (λ R)/(4 π ε0) Simplify[AsymptoticIntegrate[ℛ/(ℛ.ℛ)^(3/2), {φ′, 0, 2 π}, {r, ∞, 3}]]

Out[40]= {(R λ Sin[θ])/(2 r^2 ε0), 0, (R λ Cos[θ])/(2 r^2 ε0)}
```

Note that even though an angle variable was used, the calculation was still performed in Cartesian coordinates. This is almost always the easiest coordinate system to work in. The field is seen to be that of a point charge $Q = 2\pi a\lambda$ to lowest order in $1/r^2$.

Example 1.22 Get the field for a ring of charge (off axis) for large r to order $1/r^4$.

```
In[41]:= $Assumptions = {R ∈ Reals, R > 0, r ∈ Reals, r > 0, φ ∈ Reals, φ > 0};
         ℛ = r {Sin[θ], 0, Cos[θ]} - a {Cos[φ'], Sin[φ'], 0};
         E = (λ R)/(4 π ϵ0) Simplify[AsymptoticIntegrate[ℛ/(ℛ.ℛ)^(3/2), {φ', 0, 2 π}, {r, ∞, 4}]]
```

$$\text{Out[42]= } \left\{-\frac{R\lambda\left(9a^2 - 8r^2 + 15a^2\cos[2\theta]\right)\sin[\theta]}{16r^4\epsilon_0},\ 0,\ \frac{R\lambda\cos[\theta]\left(3a^2 + 8r^2 - 15a^2\cos[2\theta]\right)}{16r^4\epsilon_0}\right\}$$

1.5.5 Disk of Charge

For a 2-dimensional charge distribution with charge per area σ, one divides the charge into infinitesimal pieces,

$$dq = \sigma dA',$$

where $dA' = (dr')(r'd\phi)$ is the differential area in polar coordinates. A disk of charge with radius a is shown in Fig. 1.13.

Example 1.23 Calculate the field due to a disk of charge on the symmetry axis.

```
In[43]:= Clear[r′]; $Assumptions = {a > 0, z > 0};
         r = {0, 0, z}; ℛ = r - r′ { Cos[ϕ′], Sin[ϕ′], 0};
         E = σ/(4 π ϵ0) ∫_0^(2π) (∫_0^a ℛ/(ℛ.ℛ)^(3/2) r′ ⅆr′) ⅆϕ′
```

$$\text{Out[45]= } \left\{0,\ 0,\ \frac{\left(1 - \frac{z}{\sqrt{a^2+z^2}}\right)\sigma}{2\epsilon_0}\right\}$$

Example 1.24 Take the limit for an infinite disk.

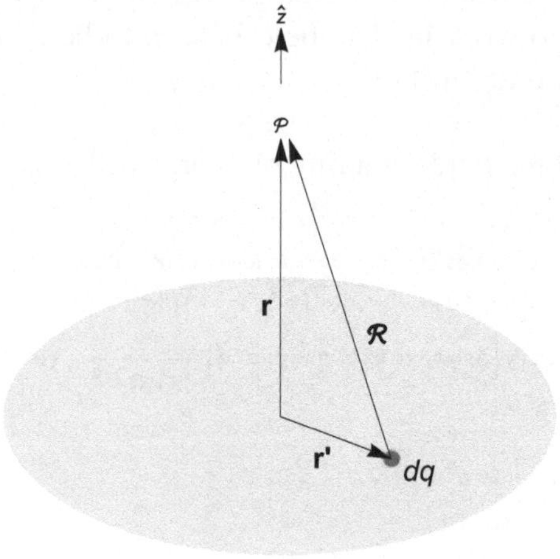

Figure 1.13 The field due to a disk of charge may be calculated by dividing it into pieces and integrating.

```
In[46]:= Limit[E, a → ∞]

Out[46]= {0, 0, σ/(2 ϵ0)}
```

The is a very important result, because any surface appears as a plane at a tiny distance. The electric field due to an infinite plane of charge is directed away (toward) from the plane for positive (negative) σ with magnitude

$$E = \frac{\sigma}{2\varepsilon_0}.$$

This result is easily calculated by the technique of Gauss's law in Chap. 2.

The ring formula may be used to get the field from the disk with $E \to dE$, $\lambda \to \sigma dr'$, and $a \to r'$ which gives

$$d\mathbf{E} = \frac{zr'\sigma dr'}{2\varepsilon_0(r'^2 + z^2)^{3/2}}\hat{\mathbf{z}}.$$

This is easily integrated to get

$$\mathbf{E} = \frac{z\sigma}{2\varepsilon_0}\int_0^a \frac{r'dr'}{(r'^2 + z^2)^{3/2}} = \frac{z\sigma}{2\varepsilon_0}\left[-(r'^2 + z^2)^{-1/2}\right]_{r'=0}^{r'=a} = \frac{\sigma}{2\varepsilon_0}\left(1 - \frac{z}{\sqrt{a^2 + z'^2}}\right).$$

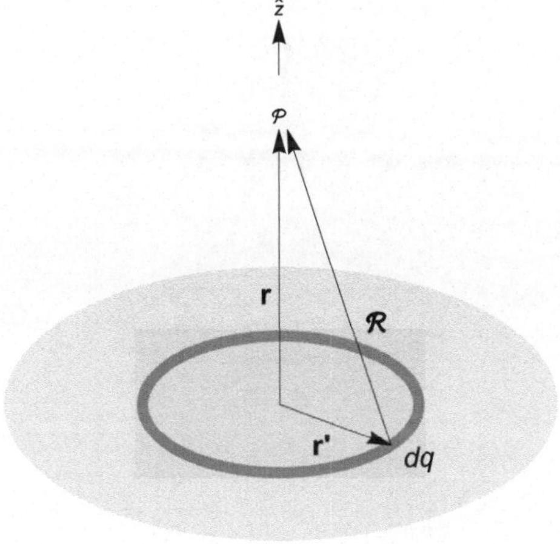

Figure 1.14 The field due to a disk of charge may be calculated by dividing it into rings and using the ring formula.

The technique of using the known field from a simpler object to get the field of another object is generally very useful.

1.5.6 Sphere of Charge

The field due to a sphere (shell) of charge as shown in Fig. 1.15 may be calculated using the ring formula with $E \to dE$, $a \to a\sin\theta'$, $z \to r - a\cos\theta'$, and $\lambda \to \sigma a d\theta'$ which gives

$$d\mathbf{E} = \frac{\sigma a^2 \sin\theta'(r - a\cos\theta')d\theta'}{2\varepsilon_0[a^2\sin^2\theta + (r - a\cos\theta')^2]^{3/2}}\hat{\mathbf{r}},$$

(Note $\hat{\mathbf{z}}$ and $\hat{\mathbf{r}}$ are the same in Fig. 1.15.) Integrating over θ',

$$\mathbf{E} = \frac{\sigma a^2}{2\varepsilon_0}\int_0^{\pi} \frac{\sin\theta'(r - a\cos\theta')d\theta'}{[a^2\sin^2\theta + (r - a\cos\theta')^2]^{3/2}}\hat{\mathbf{r}}.$$

This is a non-trivial integral. Mathematica can do a indefinite integral much faster than a definite integral. The function $f[x\_]$ is defined to be the indefinite integral, and then the limits are substituted in a second step. This is only for speed of calculation.

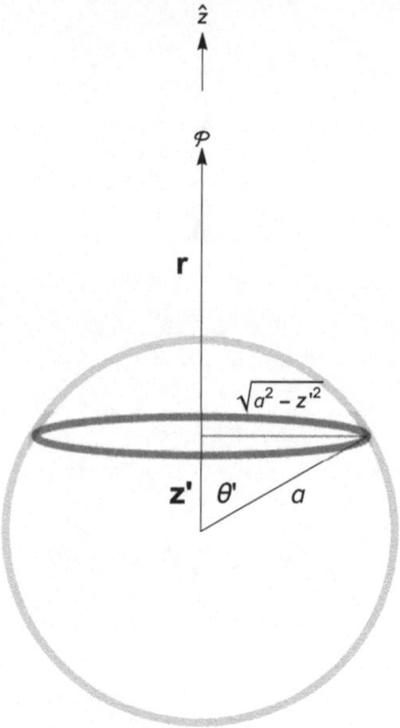

Figure 1.15 The field due to a sphere of charge may be calculated by dividing it into rings and using the ring formula.

Example 1.25 Calculate the field outside a sphere of charge by adding rings.

In[47]:=
```
Clear[f, r]; $Assumptions = {a > 0, r > a};
```
$$f[\theta\_] = \frac{\sigma a^2}{2\,\varepsilon_0}\,\text{Integrate}\left[\frac{\text{Sin}[\theta]\,(r - a\,\text{Cos}[\theta])}{\left((a\,\text{Sin}[\theta])^2 + (r - a\,\text{Cos}[\theta])^2\right)^{3/2}}, \theta\right];$$

$$\text{FullSimplify}[f[\pi] - f[0]]\ /.\ \sigma \to \frac{Q}{4\,\pi\,a^2}$$

Out[49]= $\frac{Q}{4\,\pi\,r^2\,\varepsilon_0}$

This is a remarkable result. The field outside a sphere of charge is the same as a point charge at the center of the sphere.

Example 1.26 Calculate the field inside a sphere of charge by adding rings.

```
In[50]:= $Assumptions = {a > 0, 0 < r < a};
         f[θ_] = (σ a^2)/(2 ε0) Integrate[(Sin[θ] (r - a Cos[θ]))/((a Sin[θ])^2 + (r - a Cos[θ])^2)^(3/2), θ];
         FullSimplify[f[π] - f[0] ] /. σ → Q/(4 π a^2)

Out[51]= 0
```

This is another remarkable result. The field inside a sphere of charge is zero.

1.5.7 Ball of Charge

The field due to a ball of charge as shown in Fig. 1.16 may be calculated using the disk formula with $E \rightarrow dE$, $\sigma \rightarrow \rho dz'$, $z \rightarrow r - z'$, and $a \rightarrow \sqrt{a^2 - z'}$ which gives

$$d\mathbf{E} = \frac{1}{2\varepsilon_0}\rho dz'\left(1 - \frac{r-z'}{\sqrt{a^2 - z'^2 + (r-z')^2}}\right)\hat{\mathbf{r}}.$$

Integrating over z',

$$\mathbf{E} = \frac{\rho}{2\epsilon_0}\int_{-a}^{a} dz'\left(1 - \frac{r-z'}{\sqrt{a^2 - z'^2 + (r-z')^2}}\right)\hat{\mathbf{r}}.$$

Example 1.27 Calculate the field outside a ball of charge by adding disks.

```
In[51]:= ClearAll["Global`*"];

In[52]:= $Assumptions = {z′ ∈ Reals, z′ > 0, r ∈ Reals, r > a, a > 0};
         ρ/(2 ε0) ∫_{-a}^{a} (1 - (r - z′)/Sqrt[a^2 - z′^2 + (r - z′)^2]) ⅆz′ /. ρ → Q/((4/3) π a^3)

Out[53]= Q/(4 π r^2 ε0)
```

The field outside a ball of charge behaves like all the charge is concentrated at the center.

It is already known that the field inside a sphere is zero. By superposition, the field inside a shell of any thickness is also zero. That means that the field inside the ball at distance r from the center is that due to the charge Q that is

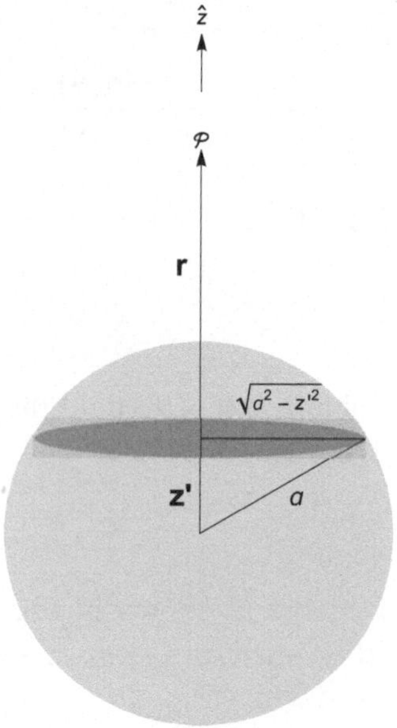

Figure 1.16 The field due to a ball of charge may be calculated by dividing it into disks and using the disk formula.

less than radius r,

$$Q = \frac{4\pi r^3}{3}\rho$$

and

$$\mathbf{E} = \frac{Q}{4\pi\varepsilon_0 r^2}\hat{\mathbf{r}} = \frac{\rho r}{3\varepsilon_0}\hat{\mathbf{r}}.$$

This can be directly checked using the disk formula, but one has to be careful about the algebraic sign of the contributions. The ball must be divided into two regions, $z' < r$ (positive contribution) and $z' > r$ (negative contribution). For the positive contribution (integral $-a$ to r) we may reverse the limits and introduce a minus sign.

Example 1.28 Calculate the field inside a ball of charge by adding disks.

```
In[54]:= $Assumptions = {z′ ∈ Reals, r > 0, r < a, a > 0};
         Simplify[-ρ/(2 ε0) ∫_r^a (1 - (z′ - r)/Sqrt[a^2 - z′^2 + (r - z′)^2]) dz′ -
           ρ/(2 ε0) ∫_r^-a (1 - (r - z′)/Sqrt[a^2 - z′^2 + (r - z′)^2]) dz′]

Out[54]= r ρ/(3 ε0)
```

1.6 AVERAGE FIELD ON A SPHERE OR BALL

Consider the field from a charge Q averaged over a sphere of arbitrary size and position. The answer depends on whether the charge is inside or outside the sphere.

Example 1.29 Calculate the average field from a point charge over a sphere of radius a with arbitrary position. Choose coordinates such that the charge is located at $(0,0,R)$ where R is an arbitrary distance.

```
In[73]:= Clear[r′]; $Assumptions = {a > 0, R > 0};
         r′ = {0, 0, R};
         ℛ = a {Sin[θ] Cos[ϕ] , Sin[θ] Sin[ϕ], Cos[θ]} - r′;
         E[θ_] = 1/(4 π a^2) Q/(4 π ε0) a^2
           Integrate[Integrate[ℛ/(ℛ.ℛ)^(3/2) Sin[θ], {ϕ, 0, 2 π}],
            θ]; E[π] - E[0] // Simplify

Out[75]= {0, 0, { -Q/(4 π R^2 ε0)   a < R
                   0                True }}
```

The field is seen to be zero if the charge is inside the sphere and

$$\mathbf{E} = -\frac{Q}{4\pi\varepsilon_0 R^2}$$

if the charge is outside the sphere.

Averaging over a ball gives the same answer as averaging over a sphere.

Example 1.30 Calculate the average field from a point charge over a ball of radius a with arbitrary position. Choose coordinates such that the charge is located at $(0,0,R)$ where R is an arbitrary distance.

```
In[37]:= Clear[r′]; $Assumptions = {a > 0, R > 0};
r′ = {0, 0, R};
ℛ = a {Sin[θ] Cos[ϕ] , Sin[θ] Sin[ϕ], Cos[θ]} - r′;
E[θ_] = 1/(4/3 π a^3) Q/(4 π ε0) r^2
  Integrate[Integrate[ℛ/(ℛ.ℛ)^(3/2) Sin[θ], {ϕ, 0, 2 π}],
   θ]; E[r_] = Integrate[E[π] - E[0], r];
E[a] - E[0] // Simplify

Out[40]= {0, 0, { -Q/(4 π R^2 ε0)   a < R
                  0                  True }}
```

1.7 CURL OF THE FIELD

For the static case, the electric field may be written in the most general case as a superposition of all charges, resulting in an integral over all space of the charge density ρ,

$$\mathbf{E}(\mathbf{r}) = \frac{1}{4\pi\varepsilon_0}\int dv' \, \frac{\rho(\mathbf{r}')\,\boldsymbol{\mathcal{R}}}{\mathcal{R}^3}.$$

The 2-dimensional case corresponds to

$$\rho dV \to \sigma dA,$$

and the 1-dimensional case corresponds to

$$\rho dV \to \lambda d\ell.$$

The electric field has an important property that its curl is zero,

$$\boldsymbol{\nabla}\times\mathbf{E} = 0.$$

The derivatives in the curl are with respect to (x,y,z), and these unprimed coordinates appear only in $\mathcal{R}$. The curl is straightforward even if there are several pieces to keep track of.

Example 1.31 Calculate the curl of $\frac{\mathcal{R}}{\mathcal{R}^3}$.

```
In[56]:= ℛ = {x, y, z} - {x′, y′, x′};
         ∇{x,y,z} × ℛ/(ℛ.ℛ)^(3/2)

Out[57]= {0, 0, 0}
```

Therefore,

$$\nabla \times \mathbf{E} = \frac{1}{4\pi\varepsilon_0} \int dv' \rho \, \nabla \times \left(\frac{\mathcal{R}}{\mathcal{R}^3} \right) = 0.$$

1.8 STRENGTH OF THE ELECTRIC FORCE

Electricity and magnetism dominates over all other forces on the human scale because it has a long range ($1/r^2$), and it is enormously stronger than gravity. The other two forces (weak and strong) have an extremely short range compared to the atomic size.

The universal gravitational constant is called with Quantity["GravitationalConstant "] and the proton mass is called with Quantity["ProtonMass"].

Example 1.32 Calculate the ratio of electric to gravitational force between 2 protons.

```
In[62]:= N[UnitConvert[(e^2/(4 π ε0))/(G mp^2)], 1]

Out[62]= 1. × 10^36
```

CHAPTER 2

Gauss's Law

The key to Gauss's law is the concept of electric flux.

2.1 ELECTRIC FLUX

The electric flux Φ through any surface is computed by defining a differential area vector

$$d\mathbf{A} = dA\hat{\mathbf{n}}.$$

where $\hat{\mathbf{n}}$ is the unit vector normal to the surface (Fig. 2.1). Notice there are 2 choices for $\hat{\mathbf{n}}$ and one of them must be chosen in order to define the sign of the flux (Φ),

$$\Phi = \int dA\, \hat{\mathbf{n}} \cdot \mathbf{E}.$$

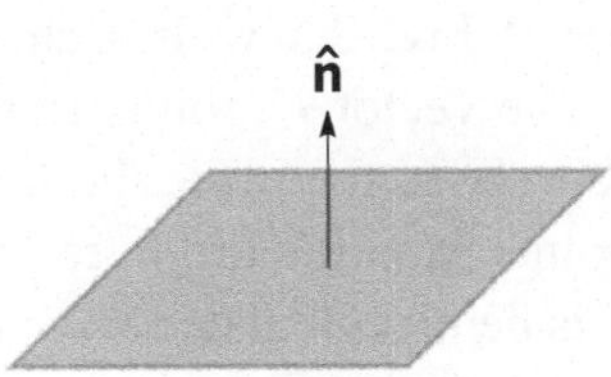

Figure 2.1 The unit normal vector is perpendicular to the surface.

The simplest case is that of a constant field such as that due to an infinite plane of charge (Ex. 1.24). Consider the flux through a surface of area A that is oriented parallel to the plane of charge as indicated in Fig. 2.2. The unit vector $\hat{\mathbf{n}}$ is taken to be in the same direction as $\mathbf{E}$ such that $\hat{\mathbf{n}} \cdot \mathbf{E} = E$. The flux is

$$\Phi = A\hat{\mathbf{n}} \cdot \mathbf{E} = EA = \frac{\sigma A}{2\varepsilon_0}.$$

The electric field is

$$E = \frac{\sigma}{2\varepsilon_0}.$$

The direction is $\hat{\mathbf{n}}$ which was known from symmetry in order to be able to evaluate the electric flux as a function E.

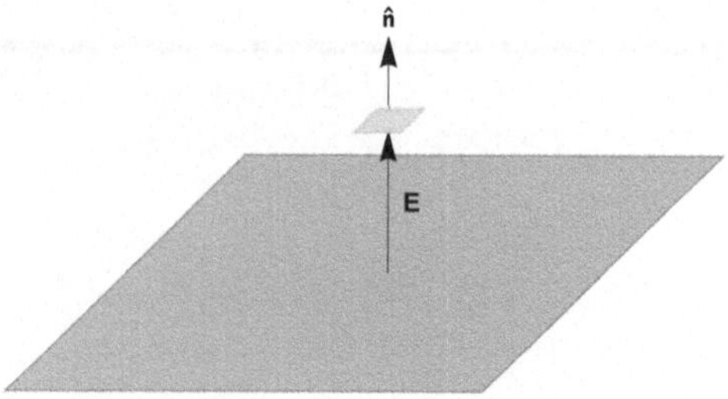

Figure 2.2 A surface element is parallel to an infinite plane of charge.

If the area element were tilted at an angle θ with respect to the original direction of $\hat{\mathbf{n}}$, the the flux becomes

$$\Phi = \int dA\, \hat{\mathbf{n}} \cdot \mathbf{E} = EA\cos\theta.$$

In general, both the magnitude and the direction of the electric field will vary on the surface and a vector integration must be performed to evaluate the flux. If there is enough symmetry, however, the vector integration may be trivial and the evaluation of the flux provides a simple equation for E. Several examples of this are shown in Sect. 2.4.

Consider the geometry of Fig. 2.3 with a charge located a distance d above a square of side L. The vector $\mathbf{r}'$ points to the charge so $r' = d$. The vector $\boldsymbol{\mathcal{R}}$ points from the charge to the place the field is being calculated on the square. The vector $\mathbf{r}$ points from the origin to an arbitrary position in the $x-y$ plane where the field is being calculated so that $\mathbf{r} = \boldsymbol{\mathcal{R}} + \mathbf{r}'$ as usual. The electric field is

$$\mathbf{E} = \frac{q}{4\pi\varepsilon_0 \mathcal{R}^3}\boldsymbol{\mathcal{R}}.$$

The flux through the square (unit normal defined to be $\hat{\mathbf{z}}$) is computed by integrating $\mathbf{E}\cdot\hat{\mathbf{n}}$ over the square,

$$\Phi = \int_{-L/2}^{L/2} dx \int_{-L2}^{L/2} dy \left(\frac{q}{4\pi\varepsilon_0 \mathcal{R}^3}\boldsymbol{\mathcal{R}}\right)\left(\frac{r'}{\mathcal{R}}\right)\cdot\hat{\mathbf{z}} = \int_{-L/2}^{L/2} dx \int_{-L2}^{L/2} dy \left(\frac{q}{4\pi\varepsilon_0 \mathcal{R}^3}\boldsymbol{\mathcal{R}}\right)\cdot\left(\frac{\mathbf{r}'}{\mathcal{R}}\right),$$

noting that $\mathbf{E}\cdot\hat{\mathbf{n}}$ picks out the z component of $\mathbf{E}$ and that $\hat{\mathbf{z}} = \mathbf{r}'/r'$.

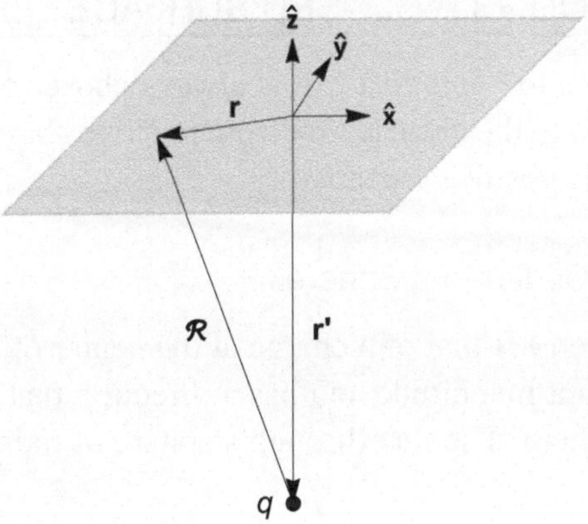

Figure 2.3 A charge q at a distance $r' = d$ along the z-axis from the center of square of side L in the $x - y$ plane.

Example 2.1 Calculate the electric flux through the square of Fig. 2.3.

```
In[1]:= $Assumptions = {z > 0, L > 0}; r′ = {0, 0, d};
        r = {x, y, 0};
        ℛ = r - r′; f[x_] = Integrate[(q ℛ/(4 π ε0 (ℛ.ℛ)^(3/2))).(r′/ℛ), x];
        g[y_] = Integrate[f[L/2] - f[-L/2], y];
        Φ = g[L/2] - g[-L/2]
```

$$\text{Out[4]=}\quad \frac{q\,\text{ArcTan}\left[\frac{L^2}{2\,d\,\sqrt{4\,d^2+2\,L^2}}\right]}{\pi\,\varepsilon_0}$$

At large distances $d >> L$, the electric field is constant on the square and the flux is just

$$\Phi = EL^2 = \frac{qL^2}{4\pi\varepsilon_0 d^2}.$$

Example 2.2 Calculate the flux for large values of distance d.

```
In[5]:= Series[Φ, {d, ∞, 3}]
```

$$\text{Out[5]=}\quad \frac{L^2\,q}{4\,\pi\,\varepsilon_0\,d^2} + O\left[\frac{1}{d}\right]^4$$

2.2 FLUX THROUGH A CLOSED SURFACE

For a closed surface, the unit vector $\hat{\mathbf{n}}$ is always chosen to be outward, such that the contribution to the electric flux from positive charges inside (outside) the closed surface is positive (negative).

2.2.1 Charge at Center of a Sphere

The simplest geometry is that of a charge at the center of a sphere. The electric field has constant magnitude and has a direction that is parallel to $\hat{\mathbf{n}}$ everywhere on the sphere. The flux through a sphere of radius a is

$$\Phi = (4\pi a^2)\left(\frac{q}{4\pi\varepsilon_0 a^2}\right) = \frac{q}{\varepsilon_0}.$$

This is a remarkable result that happens to hold true for any closed surface. The reason for this is that the electric field drops as $1/r^2$ and so does the solid angle subtended by a differential area element dA. This is often explained with the concept of electric field line. Figure 2.4 shows a plot of the vector electric field of a point charge. Every point in space gets a vector assigned with magnitude and direction of the field. The electric field lines follow the vector field directions. One draws an arbitrary number of lines (enough to see the effect but now so many as to clutter the drawing) that emanate from the charge with arrows that point in the direction of the field. The density of the lines then visually represents the relative strength and direction of the field, and the flux is proportional to the number of lines. For any closed surface that contains the charge, all the electric field lines must pass through the surface giving the same flux for a surface of any shape.

For a charge outside a closed surface, all the electric field lines that enter the surface (negative flux) must also exit the surface (positive flux), giving zero net flux (Fig. 2.5).

2.2.2 Flux Through a Hemisphere

Consider a hemisphere of radius a in a uniform electric field (Fig. 2.6). There is no charge inside it, so the flux is zero. This means that the flux through the flat end, which is

$$\Phi_{\text{flat}} = -\pi a^2 E,$$

must equal the negative of the flux though the curved part,

$$\Phi_{\text{curved}} = \pi a^2 E.$$

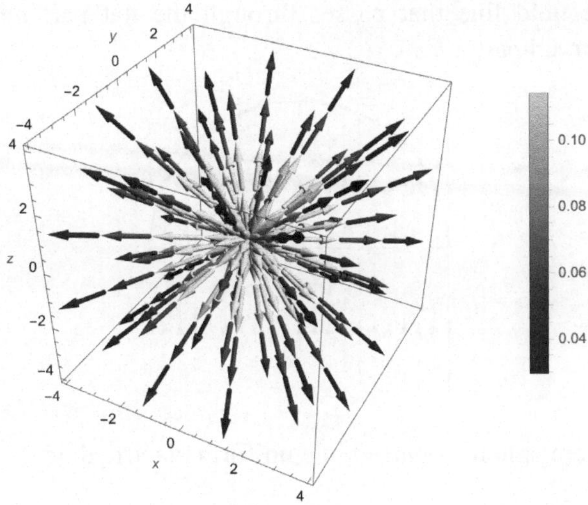

Figure 2.4 Plot of the vector electric field of a positive point charge. Electric field lines follow the vector directions and and point away from the charge. The density of the field lines visually represents the relative strength of the electric field.

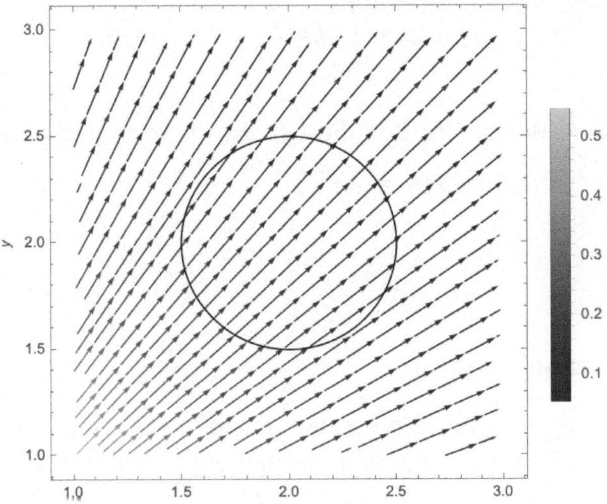

Figure 2.5 For a charge q outside a closed surface, all electric field lines that enter the surface must also exit.

Every electric field line that passes through the flat part must also pass through the curved part.

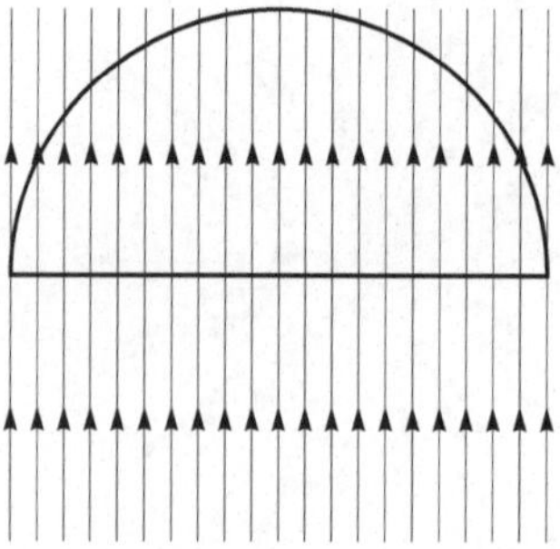

Figure 2.6 A hemisphere is placed in a uniform electric field.

Take the field to be in the z direction. For the round part,

$$\mathbf{E} \cdot \hat{\mathbf{n}} = E \cos\theta,$$

where θ is the polar angle (the angle measured from the z-axis). The flux through the round part is

$$\Phi = 2\pi a \int_{-\pi/2}^{\pi/2} d\theta \, (E\cos\theta)(a \sin\theta).$$

Example 2.3 Calculate the flux through the curved part of the hemisphere.

In[6]:= $2\,\pi\,a \int_0^{\frac{\pi}{2}} \mathrm{E}\,\mathrm{Cos}[\theta]\;a\,\mathrm{Sin}[\theta]\,d\theta$

Out[6]= $a^2\,\pi\,\mathrm{E}$

2.2.3 Flux Through a Cube

Consider a charge placed at the center of a cube with side L (Fig. 2.7). In Ex. 2.1, for $d \to L/2$, one gets the flux through one side of a cube.

Example 2.4 Calculate the flux in Ex. 2.1 for $d \to L/2$,

In[7]:= $\mathrm{Limit}\left[\Phi,\; d \to \frac{L}{2}\right]$

Out[7]= $\frac{q}{6\,\varepsilon_0}$

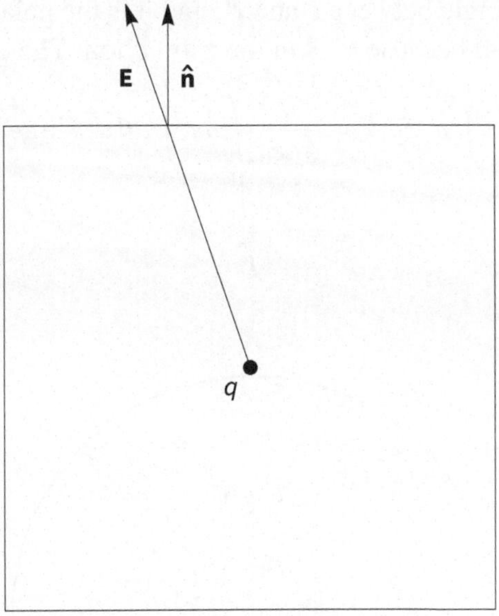

Figure 2.7 A charge q is placed iat the center if a cube.

The flux though the entire cube from a charge at the center is

$$\Phi = (6)\left(\frac{q}{6\varepsilon_0}\right) = \frac{q}{\varepsilon_0}.$$

2.2.4 Flux Through a Sphere from a Charge Inside

Figure 2.8 shows a charge q placed inside some spherical boundary of radius a at an arbitrary position. The origin is chosen to be at the center of the sphere, and the z direction is chosen to be along a line passing through q. The vector $\mathbf{r}'$ points to the charge. The vector $\mathbf{r}$ points to the sphere and the vector $\boldsymbol{\mathcal{R}}$ points from the charge to the sphere. At the arbitrary position $\mathbf{r}$, the electric field is

$$\mathbf{E} = \frac{q}{4\pi\varepsilon_0 \mathcal{R}^3}\boldsymbol{\mathcal{R}}.$$

The unit vector normal to the sphere is

$$\hat{\mathbf{n}} = \hat{\mathbf{r}}.$$

The angle θ is the angle between $\mathbf{r}$ and $\mathbf{r}'$, and it is the polar angle (angle with respect to the z-axis) because $\mathbf{r}'$ is in the z direction. The electric flux is

$$\Phi = \int dA\hat{\mathbf{n}} \cdot \mathbf{E} = \frac{q}{4\pi\varepsilon_0} \int_0^{2\pi} d\phi \int_0^{\pi} d\theta\, a^2 \sin\theta \frac{\hat{\mathbf{r}} \cdot \mathbf{E}}{\mathcal{R}^3}.$$

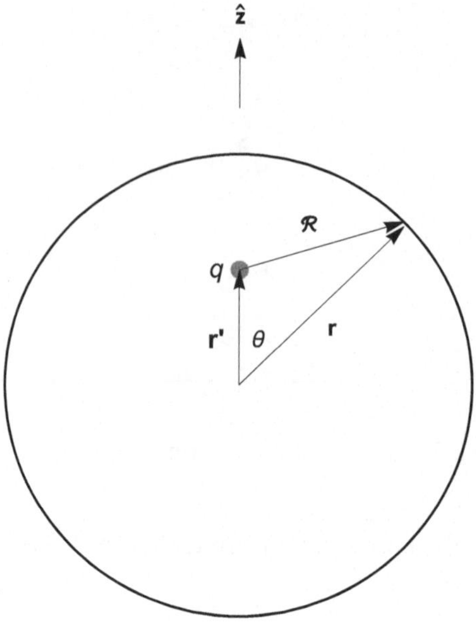

Figure 2.8 A charge q is placed at an arbitrary position inside an arbitrary spherical boundary.

Example 2.5 Calculate the electric flux through the sphere of Fig. 2.8 where the charge is inside.

```
In[8]:= $Assumptions = {0 < z < a}; r′ = {0, 0, z};
        r = a {Sin[θ] Cos[ϕ], Sin[θ] Sin[ϕ], Cos[θ]};
        ℛ = r - r′;
        f[θ_] =
         Integrate[Integrate[(q ℛ / (4 π ε0 (ℛ.ℛ)^(3/2))).(r/a) a^2 Sin[θ],
           {ϕ, 0, 2 π}], θ];
        Simplify[f[π] - f[0]]

Out[10]= q/ε0
```

2.2.5 Flux Through a Sphere from a Charge Outside

A charge outside the spherical boundary (Fig. 2.9) has a net zero flux through the sphere. On one section of the sphere $\mathbf{E} \cdot \hat{\mathbf{n}}$ is negative, while on another part $\mathbf{E} \cdot \hat{\mathbf{n}}$ is positive.

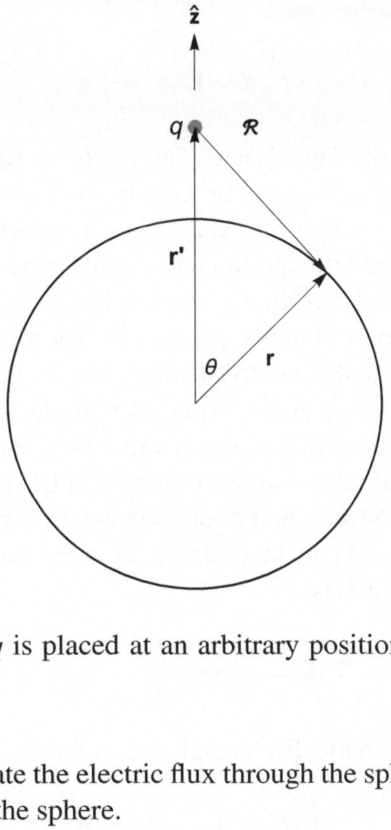

Figure 2.9 A charge q is placed at an arbitrary position outside an arbitrary spherical boundary.

Example 2.6 Calculate the electric flux through the sphere of Fig. 2.9 where the charge is outside the sphere.

```
In[11]:= $Assumptions = {z > a > 0}; r′ = {0, 0, z};
         r = a {Sin[θ] Cos[ϕ], Sin[θ] Sin[ϕ], Cos[θ]};
         𝓡 = r - r′;
         f[θ_] =
          Integrate[Integrate[(q 𝓡 / (4 π ε0 (𝓡.𝓡)^(3/2))).(r/a) a^2 Sin[θ],
            {ϕ, 0, 2 π}], θ];
         Simplify[f[π] - f[0]]

Out[13]= 0
```

2.3 MAXWELL EQUATION

So far, only a single charge has been considered. By superposition, any charge q_i inside contributes to the flux with q_i/ε_0, and any charge outside contributes zero. The electric flux through any closed surface is equal to the net enclosed charge (q_{en}) divided by ε_0.

$$\oint d\mathbf{A} \cdot \mathbf{E} = \frac{q_{\text{en}}}{\varepsilon_0}.$$

This is the statement of Gauss's law. The circle on top of the normal integral sign is the standard notation for a an integral over a closed surface. It is understood that the unit normal vector in $d\mathbf{A} = dA\hat{\mathbf{n}}$ is outward from the surface. It is a remarkable (experimental) result. One should take note that the electric field for a moving charge is complicated and that Coulomb's law does not hold when charges move close to the speed of light. Nevertheless, Gauss's law still holds and is relativistically correct. The field due to moving charges is the subject of an advanced course on electromagnetism. Gauss's law is one of the four Maxwell equations that describe all of classical electromagnetism. The other three are encountered in Chaps. 4, 7, and 11. While it is sometimes not used in a first course on electricity and magnetism, it is probably good at least to know that this equation can be written in differential form. The divergence of $\mathbf{E}$ is

$$\nabla \cdot \mathbf{E} = \frac{\partial E_x}{\partial x} + \frac{\partial E_y}{\partial y} + \frac{\partial E_z}{\partial z}.$$

It can be proven mathematically, a result called the divergence theorem, that

$$\int dv\, \nabla \cdot \mathbf{E} = \oint d\mathbf{a} \cdot \mathbf{E},$$

where the integral on the right is over the closed surface that surrounds the volume that is integrated on the left. An example of the divergence theorem is given in App. B.5. Using

$$q_{\text{en}} = \int dv\, \rho,$$

where ρ is the charge density, one arrives at

$$\nabla \cdot \mathbf{E} = \frac{\rho}{\varepsilon_0}.$$

This is Gauss's law in differential form. The use of the differential form requires fluency in multivariable vector calculus and is typically reserved for a higher level course.

The physical interpretation of Gauss's law is that charge is the source of electric field, the field diverges exactly at the place where charge is located, and the field has zero divergence everywhere else. In higher mathematics, the infinite divergence at the location of a point charge is described with the Dirac delta function, $\delta(\mathbf{r})$. The Dirac delta function has the property that it is zero everywhere except where its argument is zero, where it comes infinite in a way that its integral is unity. The charge density of a point charge is

$$\rho = q\delta(\mathbf{r}),$$

where $\mathbf{r}$ is the location of the charge. The integral is

$$\int dv\, \rho = q \int dv \delta(\mathbf{r}) = q.$$

The electric field of the point charge is

$$\mathbf{E} = \frac{q}{4\pi\varepsilon_0 r^2}\hat{\mathbf{r}}.$$

Gauss's law says

$$\nabla \cdot \mathbf{E} = \frac{q}{4\pi\varepsilon_0}\nabla \cdot \frac{\hat{\mathbf{r}}}{r^2} = \frac{q\delta(\mathbf{r})}{\varepsilon_0}$$

This means that

$$\nabla \cdot \frac{\hat{\mathbf{r}}}{r^2} = 4\pi\delta(\mathbf{r}).$$

The divergence of the electric field due to a point charge is zero everywhere except at the location of the point charge where it is infinite.

2.4 APPLYING GAUSS'S LAW

To use Gauss's law to calculate the field, there must be symmetry such that the direction of the field is known. One then draws a simple closed surface according to the symmetry, and then calculates the flux using the magnitude of the field E as a variable. The flux is set equal to the enclosed charge divided by ε_0 giving an equation for E.

2.4.1 Line of Charge

Consider a line of charge with constant charge density λ. The electric field is cylindrically symmetric (has no ϕ dependence) and varies with radial distance r. For the closed surface, one chooses a concentric cylinder with radius r and length L (Fig. 2.10. Note that the choice of L does not matter as it will cancel and the arbitrary distance r is the place where the field is evaluated. The key features of the chosen surface is that the field is constant on the curved part of the cylinder and in the same direction as the outward normal $\hat{\mathbf{n}}$ and that on the flat ends $\mathbf{E} \cdot \hat{\mathbf{n}} = 0$.

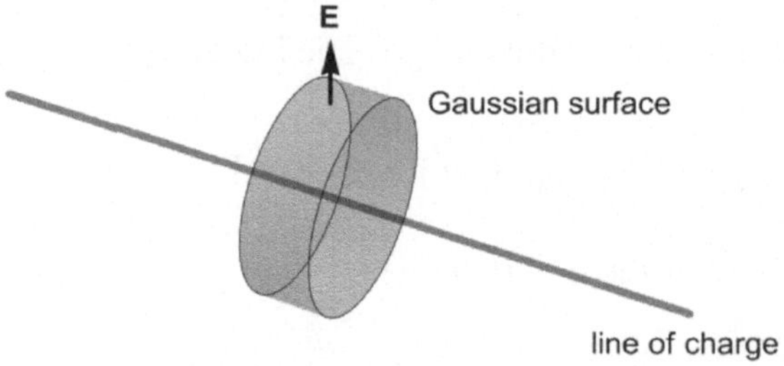

Figure 2.10 For a line of charge, a concentriic cylindrical gausian surface is chosen.

The flux is

$$\Phi = \oint d\mathbf{A} \cdot \mathbf{E} = 2\pi r L E$$

and by Gauss's law,

$$\Phi = 2\pi r L E = \frac{q_{\text{en}}}{\varepsilon_0} = \frac{\lambda L}{\varepsilon_0}.$$

Solve for E,

$$E = \frac{\lambda}{2\pi\varepsilon_0 r}.$$

and

$$\mathbf{E} = \frac{\lambda}{2\pi\varepsilon_0 r}\hat{\mathbf{r}},$$

This is the same result as obtained by direct integration using Coulomb's law in Ex. 1.15.

2.4.2 Plane of Charge

For an infinite plane of charge with constant charge density σ, the electric field is normal to the plane by symmetry. For a closed surface, one chooses what has come to be known as a "Gaussian pillbox," which is a tiny container with its flat surfaces (area A) placed parallel and equal distance to the plane.Key features of this surface are that $\mathbf{E}$ is constant on the flat part and in the same direction as the outward normal $\hat{\mathbf{n}}$ and that $\mathbf{E}\cdot\hat{\mathbf{n}} = 0$ on the rest of the surface. It does not matter if the box has a round or square cross-section.

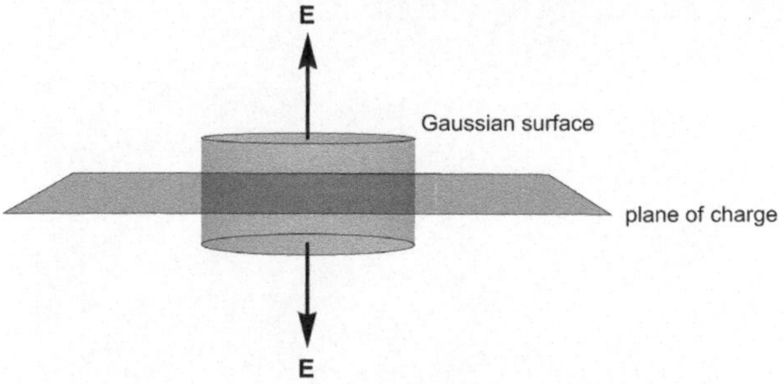

Figure 2.11 For a plane of charge, a Gaussian surface is chosen with flat pieces that are parallel to the plane.

The flux is

$$\Phi = \oint d\mathbf{A}\cdot\mathbf{E} = 2AE$$

and by Gauss's law,

$$\Phi = 2AE = \frac{q_{\text{en}}}{\varepsilon_0} = \frac{\sigma A}{\varepsilon_0}.$$

Solve for E,

$$E = \frac{\sigma}{2\varepsilon_0},$$

and if the charge is in the $x-y$ plane,

$$\mathbf{E} = \frac{\sigma}{2\varepsilon_0}\hat{\mathbf{z}},$$

for positive z. Note that the field points away from the plane of charge if the charge is positive. This is the same result as obtained by direct integration using Coulomb's law in Ex. 1.24.

2.4.3 Ball of Charge

For a ball of charge of radius a, one chooses a spherical Gaussian surface oriented such that the electric field is everywhere perpendicular to the surface.

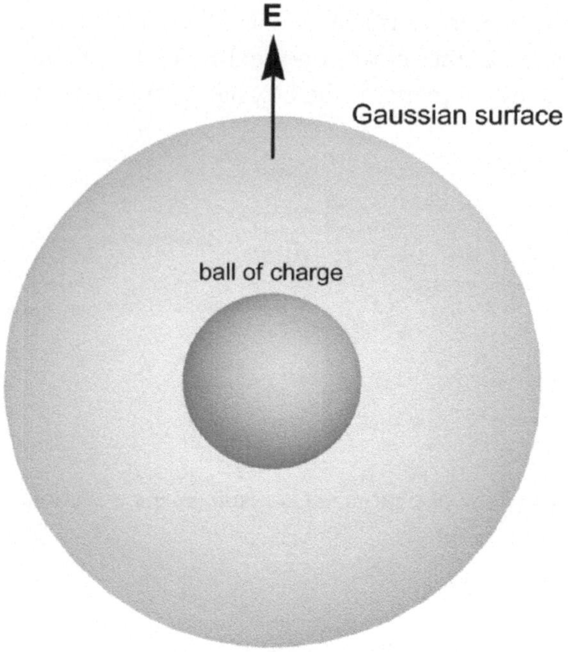

Figure 2.12 For a ball charge, a gausian surface is chosen to be a concentric sphere

Outside the ball, (Fig. 2.12) $r > a$. The flux is

$$\Phi = \oint d\mathbf{A} \cdot \mathbf{E} = 4\pi r^2 E$$

and by Gauss's law,

$$\Phi = 4\pi r^2 E = \frac{q_{\text{en}}}{\varepsilon_0} = \frac{4\pi a^3 \rho}{3\varepsilon_0}.$$

Solve for E,

$$E = \frac{a^3 \rho}{3\varepsilon_0 r^2},$$

and

$$\mathbf{E} = \frac{a^3\rho}{3\varepsilon_0 r^2}\hat{\mathbf{r}} = \frac{Q}{4\pi\varepsilon_0 r^2}\hat{\mathbf{r}},$$

where $Q = \frac{4}{3}\pi a^3$ is the total charge of the ball. This is the same result as obtained by direct integration using Coulomb's law in Ex. 1.27.

Inside the ball (Fig. 2.13), $r < a$.

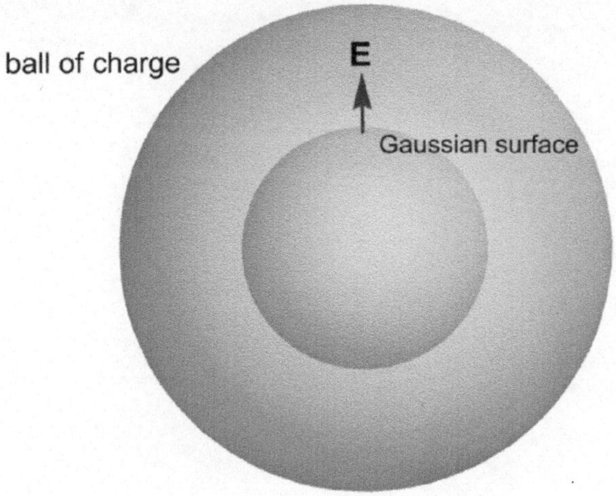

Figure 2.13 For a ball charge, a Gaussian surface is chosen to be a concentric sphere.

The flux is the same as before (r is just a parameter)

$$\Phi = \oint d\mathbf{A} \cdot \mathbf{E} = 4\pi r^2 E$$

but now only a fraction of the charge is enclosed, with Gauss's law giving

$$\Phi = 4\pi r^2 E = \frac{q_{\text{en}}}{\varepsilon_0} = \frac{4\pi r^3 \rho}{3\varepsilon_0}.$$

Solve for E,

$$E = \frac{\rho r}{3\varepsilon_0},$$

and

$$\mathbf{E} = \frac{\rho r}{3\varepsilon_0}\hat{\mathbf{r}}.$$

This is the same result as that obtained by direct integration using Coulomb's law in Ex. 1.28.

CHAPTER 3

Electric Potential

The electric potential (V) is the potential energy (U) per charge. Only changes in potential energy (ΔU) have physical meaning and the same is true for electric potential differences (ΔV),

$$\Delta V = \frac{\Delta U}{q}.$$

The unit of the electric potential is J/C which is the definition of volt (V).

3.1 ELECTRIC POTENTIAL DIFFERENCE

The electric force is conservative, which means that the work (W) done in moving a charge q from $\mathcal{A}$ to $\mathcal{B}$ does not depend on its path. The change in potential energy is

$$\Delta U = U_{\mathcal{B}} - U_{\mathcal{A}} = -W = -\int_{\mathcal{A}}^{\mathcal{B}} d\boldsymbol{\ell} \cdot \mathbf{F} = -q \int_{\mathcal{A}}^{\mathcal{B}} d\boldsymbol{\ell} \cdot \mathbf{E},$$

and

$$\Delta V = \frac{\Delta U}{q} = -\int_{\mathcal{A}}^{\mathcal{B}} d\boldsymbol{\ell} \cdot \mathbf{E},$$

3.2 CURL OF E

The source of electric field is electric charge, and for a single charge the field varies as $1/r^2$. The divergence has a singularity because the field depends on r in the direction $\hat{\mathbf{r}}$. The field does not have components along $\hat{\boldsymbol{\theta}}$ or $\hat{\boldsymbol{\phi}}$, so the curl does not have such a singularity.

Example 3.1 Calculate the curl of $1/r^2$ in spherical coordinates.

```
In[1]:= Curl[{1/r^2, 0, 0}, {r, θ, ϕ}, "Spherical"]

Out[1]= {0, 0, 0}
```

Example 3.2 Calculate the curl of $1/r^2$ in Cartesian coordinates.

```
In[2]:= Curl[{x, y, z}/(x^2 + y^2 + z^2)^(3/2), {x, y, z}]

Out[2]= {0, 0, 0}
```

Mathematically, it is the zero curl of the static electric field that allows us to write

$$\int d\mathbf{a} \cdot (\nabla \times \mathbf{E}) = 0 = \oint d\boldsymbol{\ell} \cdot \mathbf{E},$$

where the last step is from Stokes' theorem. An example of Stokes' theorem is given in App. B.6. For any path that ends where it starts, the initial and final potential must be identical.

3.3 POINT CHARGE

For a point charge,

$$\Delta V = -\int_{\mathcal{A}}^{\mathcal{B}} d\boldsymbol{\ell} \cdot \mathbf{E} = -\frac{q}{4\pi\varepsilon_0} \int_{\mathcal{A}}^{\mathcal{B}} d\boldsymbol{\ell} \cdot \left(\frac{\hat{\mathbf{r}}}{r^2}\right) = \frac{q}{4\pi\varepsilon_0} \left(\frac{1}{r_{\mathcal{B}}} - \frac{1}{r_{\mathcal{A}}}\right).$$

If the zero of the potential is placed at infinity, then the potential at any value of r is

$$V = \frac{q}{4\pi\varepsilon_0 r}.$$

3.3.1 Field from the Potential

The electric field may be obtained from the electric potential by taking its negative derivative. In one dimension, or with sufficient symmetry, this is a simple derivative, and the direction of the field is along the coordinate that is being differentiated. For a point charge

$$\mathbf{E} = -\frac{d}{dr}\left(\frac{q}{4\pi\varepsilon_0 r}\right)\hat{\mathbf{r}} = \frac{q}{4\pi\varepsilon_0 r^2}\hat{\mathbf{r}}.$$

One needs to be careful here. The derivative can only be calculated this way because the field is only in the r direction, and there is no angle dependence.

In three dimensions the derivative becomes the gradient. In Cartesian coordinates,

$$\mathbf{E} = -\nabla V = -\frac{dV}{dx}\hat{\mathbf{x}} + -\frac{dV}{dy}\hat{\mathbf{y}} + -\frac{dV}{dz}\hat{\mathbf{z}}.$$

For the point charge,

$$E_x = -\frac{d}{dx}\left(\frac{q}{4\pi\epsilon_0\sqrt{x^2+y^2+z^2}}\right) = \frac{qx}{4\pi\varepsilon_0(x^2+y^2+z^2)^{3/2}},$$

$$E_y = -\frac{d}{dy}\left(\frac{q}{4\pi\epsilon_0\sqrt{x^2+y^2+z^2}}\right) = \frac{qy}{4\pi\varepsilon_0(x^2+y^2+z^2)^{3/2}},$$

and

$$E_z = -\frac{d}{dz}\left(\frac{q}{4\pi\epsilon_0\sqrt{x^2+y^2+z^2}}\right) = \frac{qz}{4\pi\varepsilon_0(x^2+y^2+z^2)^{3/2}}.$$

The field is

$$\mathbf{E} = \frac{q(x\hat{\mathbf{x}} + y\hat{\mathbf{y}} + z\hat{\mathbf{z}})}{4\pi\varepsilon_0(x^2+y^2+z^2)^{3/2}},$$

which is exactly what is meant by the spherical-coordinate form

$$\mathbf{E} = \frac{q\hat{\mathbf{r}}}{4\pi\varepsilon_0 r^3}.$$

This is a very simple example, but the following code can be used as a template to calculate the field for a more complicated example. The function Grad[f, $\{x_1, x_2, x_3\}$] takes the gradient in the specified coordinate system, and TransformedField[t, f, $\{x_1, x_2, x_3\} \to \{y_1, y_2, y_3\}$] transforms from one coordinate system to another.

Example 3.3 Calculate the electric field for a point charge from the electric potential in Cartesian coordinates and transform the result into spherical coordinates.

```
In[3]:= $Assumptions = r > 0;
        V = q/(4 π ε0 Sqrt[x^2 + y^2 + z^2]); E = -Grad[V, {x, y, z}];
        TransformedField["Cartesian" → "Spherical", E,
          {x, y, z} → {r, θ, ϕ}] // Simplify

Out[4]= { q/(4 π r^2 ε0), 0, 0}
```

The definition of potential leads to a more intuitive unit for the electric field, V/m. The electric field unit is

$$\frac{\text{N}}{\text{C}} = \frac{\text{J/m}}{\text{C}} = \frac{\text{J/C}}{\text{m}} = \frac{\text{V}}{\text{m}}.$$

3.4 FORMULA FOR V

The integral expression for **E** together with $\mathbf{E} = -\nabla V$ allows us to write V relative to infinity for a line of charge with density λ as

$$V = \frac{1}{4\pi\varepsilon_0} \int d\ell' \frac{\lambda}{\mathcal{R}},$$

for a surface charge of density σ as

$$V = \frac{1}{4\pi\varepsilon_0} \int da' \frac{\rho}{\mathcal{R}},$$

and for a volume charge of density ρ as

$$V = \frac{1}{4\pi\varepsilon_0} \int dv' \frac{\rho}{\mathcal{R}},$$

These formulae follow from the general formula for **E** (Sect. 1.5.2).

3.5 LINE OF CHARGE

Consider a line of charge with uniform density λ as shown in Fig. 3.1. The potential may be calculated with code very similar to Ex. 1.13, except the integration is now scalar $1/r$ instead of vector $\hat{\mathbf{r}}/r^2$.

Example 3.4 Calculate the electric potential (relative to infinity) along the midpoint of the line of charge of Fig. 3.1.

```
In[5]:= ClearAll["Global`*"];
        $Assumptions = {L > 0, y > 0};
        r = {0, y, 0}; r′ = {x′, 0, 0}; ℛ = r - r′;
```

$$\texttt{V[y\_]} = \frac{\lambda}{4\pi\varepsilon_0}\int_{-L/2}^{L/2} \frac{1}{(\mathcal{R}.\mathcal{R})^{1/2}}\, dx'$$

$$\text{Out[7]=}\ \frac{\lambda \,\text{Log}\left[1 + \frac{L\left(L+\sqrt{L^2+4y^2}\right)}{2y^2}\right]}{4\pi\varepsilon_0}$$

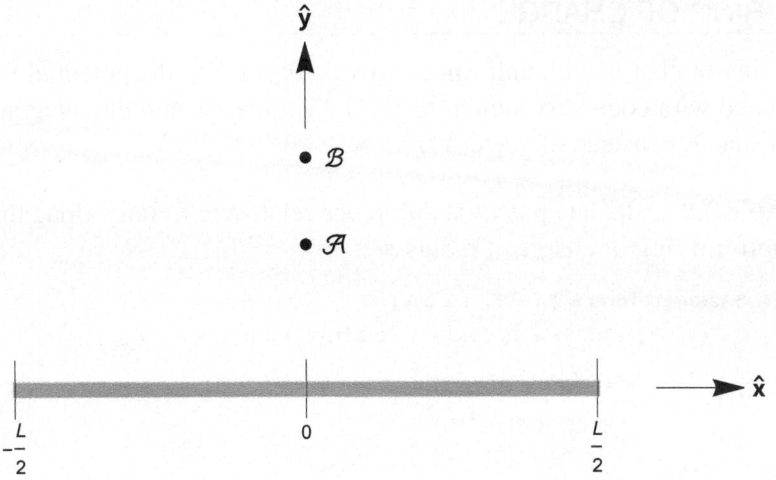

Figure 3.1 A line of charge with constant density λ extends from $-L/2 < x < L/2$.

The electric field for an infinite line of charge was calculated in 1.5.3 to be

$$\mathbf{E} = \frac{\lambda}{2\pi\epsilon_0 r}\hat{\mathbf{r}}.$$

The potential difference between point $\mathcal{A}$ and point $\mathcal{B}$ can be caculated with

$$\Delta V = -\int_{\mathcal{A}}^{\mathcal{B}} d\boldsymbol{\ell}\cdot\mathbf{E} = -\frac{\lambda}{2\pi\epsilon_0}\int_{\mathcal{A}}^{\mathcal{B}} dr\frac{\hat{\mathbf{r}}\cdot\hat{\mathbf{r}}}{r} = \frac{\lambda}{2\pi\epsilon_0}\ln\left(\frac{r_{\mathcal{A}}}{r_{\mathcal{B}}}\right).$$

Example 3.5 Calculate the potential difference between point $\mathcal{A}$ and point $\mathcal{B}$ of Fig. 3.1 for $\mathcal{A} \to y_1$ and $\mathcal{B} \to y_2$.

```
In[8]:= << Notation`
        Symbolize[ParsedBoxWrapper[SubscriptBox["_", "_"]]]
        $Assumptions = {y₁ > 0, y₂ > 0};
        Limit[V[y₂] - V[y₁], L → ∞]
```

Out[11]= $\dfrac{\lambda \,\mathrm{Log}\left[\frac{y_1}{y_2}\right]}{2\,\pi\,\varepsilon_0}$

Note that point $\mathcal{A}$ is at a higher potential than point $\mathcal{B}$.

3.6 RING OF CHARGE

For a ring of charge with uniform density λ (Fig. 1.12), the potential may be calculated with code very similar to Ex. 1.19, except again the integration is now scalar $1/r$ instead of vector $\hat{\mathbf{r}}/r^2$.

Example 3.6 Calculate potential difference relative to infinity along the axis of a uniform ring of charge of radius a.

```
In[12]:= $Assumptions = {a > 0, z ∈ ℝ};
         r = {0, 0, z}; r′ = {a Cos[ϕ′], a Sin[ϕ′], 0}; ℛ = r - r′;
                     2π
                 λ a ⌠        1
         V[z_] = ─── │ ─────────── ⅆϕ′
                 4πε0 ⌡ (ℛ.ℛ)^(1/2)
                     0
```

```
                a λ
Out[14]= ─────────────────
          2 √(a² + z²) ε0
```

This expression for the electric potential can be reproduced by integration of the electric field that was determined in Ex. 1.19,

$$\mathbf{E} = \frac{az\lambda}{2\varepsilon_0(a^2+z^2)^{3/2}}\hat{\mathbf{z}},$$

which gives

$$\Delta V = -\int_z^\infty d\boldsymbol{\ell}\cdot\mathbf{E} = -\frac{a\lambda}{2\varepsilon_0}\int_z^\infty dz'\frac{z'}{(a^2+z'^2)^{3/2}} = \frac{a\lambda}{2\varepsilon_0\sqrt{a^2+z^2}}.$$

3.7 DISK OF CHARGE

The electric potential at a distance z on the symmetry axis of a disk of charge (Fig. 1.13) with uniform charge density σ may be calculated by adding up rings, where λ in the ring formula is replaced by $\sigma dr'$ and the radius of the ring a is replaced by r'. A differential piece of the potential is

$$dV = \frac{r'\sigma dr'}{2\varepsilon_0\sqrt{r'^2+z^2}}.$$

Example 3.7 Calculate potential difference relative to infinity along the axis of a uniform disk of charge of radius a.

```
In[15]:= ClearAll["Global`*"];
         $Assumptions = {z > 0, a > 0} ;
```

$$\frac{\sigma}{2\,\varepsilon_0}\int_0^a \frac{r\prime}{\left(z^2 + r\prime^2\right)^{1/2}}\,dr\prime$$

$$\text{Out[15]=}\ \frac{\left(-z + \sqrt{a^2 + z^2}\right)\sigma}{2\,\varepsilon_0}$$

This expression for the electric potential can be reproduced by integration of the electric field that was determined in 1.19,

$$\mathbf{E} = \frac{\sigma}{2\varepsilon_0}\left(1 - \frac{z}{\sqrt{a^2 + z^2}}\right)\hat{\mathbf{z}},$$

which gives

$$\Delta V = -\int_z^\infty d\boldsymbol{\ell}\cdot\mathbf{E} = -\frac{\sigma}{2\varepsilon_0}\int_z^\infty dz'\left(1 - \frac{z}{\sqrt{a^2 + z^2}}\right) = \frac{\sigma}{2\varepsilon_0}\left(-z + \sqrt{a^2 + z^2}\right).$$

3.8 SPHERE OF CHARGE

The potential outside a sphere of charge of radius a must be the same as a point charge because the electric field is the same as a point charge. Nonetheless, this may be easily checked by adding up rings (Fig. 3.2). The radius of a ring is $a\sin\theta$, and the distance to the center is $r - a\cos\theta$.

Example 3.8 Calculate the potential difference outside a sphere with uniform surface charge density σ.

```
In[16]:= ClearAll["Global`*"];
         $Assumptions = {r > a, a > 0};
         f[θ_] =
```

$$\frac{\sigma\, a^2}{2\,\varepsilon_0}\,\text{Integrate}\left[\frac{\text{Sin}[\theta]}{\left((a\,\text{Sin}[\theta])^2 + (r - a\,\text{Cos}[\theta])^2\right)^{1/2}},\ \theta\right];$$

$$\text{FullSimplify}[f[\pi] - f[0]]\ /.\ \sigma \to \frac{Q}{4\,\pi\,a^2}$$

$$\text{Out[17]=}\ \frac{Q}{4\,\pi\,r\,\varepsilon_0}$$

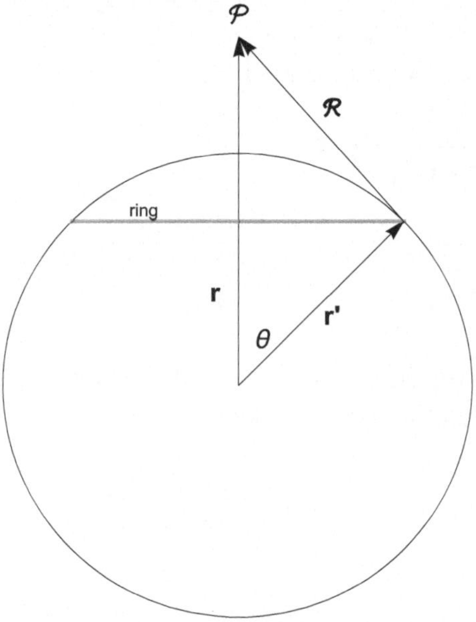

Figure 3.2 The electric potential due to a sphere of charge may be calculated by dividing it into rings and using the ring formula.

Inside the sphere of charge, the electric field is zero, and the potential does not change. It has the same value as it does on the surface (radius a),

$$V = \frac{Q}{4\pi\varepsilon_0 a}.$$

3.9 BALL OF CHARGE

The potential outside a uniform ball of charge of radius a is also identical to that of a point charge. We may calculate the potential due to a ball by adding up disks using the result of Sect. 3.7. The charge density of each disk (Fig. 1.16) is

$$\sigma = \rho dz',$$

the distance to the center is

$$r - z',$$

and the radius is

$$\sqrt{a^2 - z'^2}.$$

Example 3.9 Calculate the potential difference outside a ball with uniform volume charge density ρ.

```
In[18]:= $Assumptions = {z' ∈ Reals, z' > 0, r ∈ Reals,
            r > a, a > 0};
```

$$\frac{\rho}{2\,\varepsilon_0}\int_{-a}^{a}\left(-(r-z')+\sqrt{a^2-z'^2+(r-z')^2}\right)dz'\ /.\ \rho\to\frac{Q}{(4/3)\,\pi\,a^3}$$

Out[19]= $\frac{Q}{4\,\pi\,r\,\varepsilon_0}$

3.10 ELECTRIC DIPOLE

A physical electric dipole consists of two charges q and $-q$ separated by a distance d (Fig. 3.3). The electric dipole is the most important example in electricity and magnetism. The reason for this is that all matter is made of atoms, and when atoms get stretched in electric fields, they form dipoles (see Fig. 3.4).

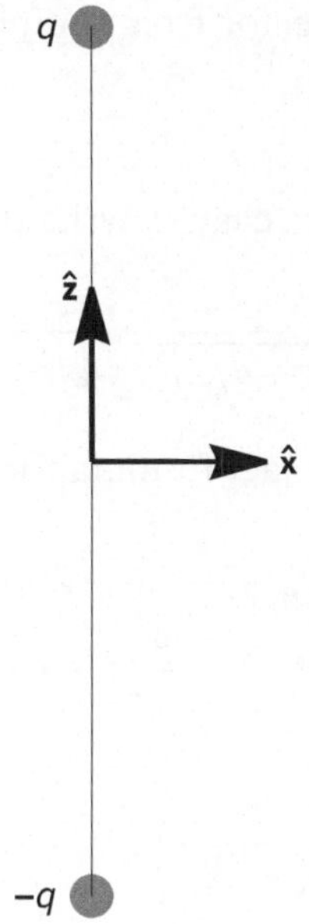

Figure 3.3 Two charges q and $-q$ separated by a distance d along the z axis form an electric dipole.

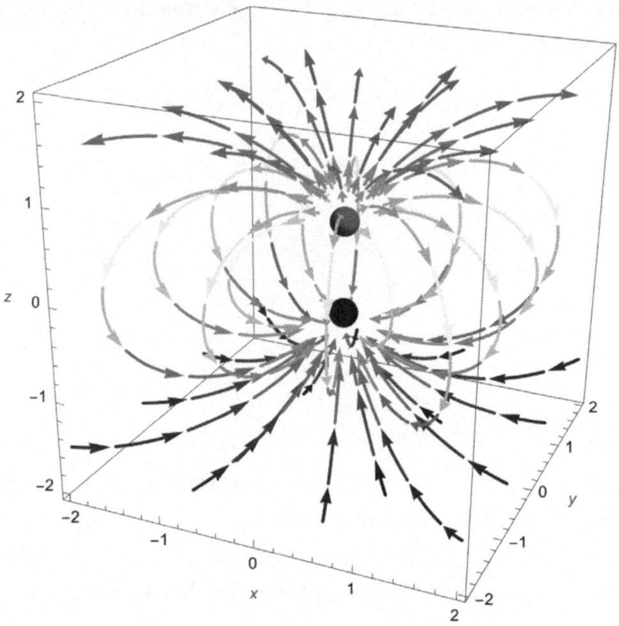

Figure 3.4 The electric field vector from a dipole makes loops that go from the positive to negative charge.

The potential for a dipole is easy to write down,

$$V = \frac{q}{4\pi\varepsilon_0\sqrt{x^2+y^2+(z+d/2)^2}} - \frac{q}{4\pi\varepsilon_0\sqrt{x^2+y^2+(z-d/2)^2}}.$$

Example 3.10 Calculate the exact formula for the dipole potential.

```
In[20]:= << Notation`
Symbolize[ParsedBoxWrapper[SubscriptBox["_", "_"]]]
r1 = {0, 0, d/2}; r2 = {0, 0, -d/2}; r = {x, y, z};
ℛ1 = r - r1; ℛ2 = r - r2;
V = q/(4 π ε0) (1/Sqrt[ℛ1.ℛ1] - 1/Sqrt[ℛ2.ℛ2])
```

$$\text{Out[22]=}\quad \frac{q\left(\frac{1}{\sqrt{x^2+y^2+\left(-\frac{d}{2}+z\right)^2}} - \frac{1}{\sqrt{x^2+y^2+\left(\frac{d}{2}+z\right)^2}}\right)}{4\pi\varepsilon_0}$$

One is interested in the limit where the separation distance is small compared to the observation distance. This is referred to as an ideal dipole. A bit of algebra is acquired to deduce the formula.

Example 3.11 Calculate the dipole potential to leading order in *d* (the ideal dipole).

```
In[23]:= Series[V, {d, 0, 2}]
```

Out[23]= $\frac{q\,z\,d}{4\,\pi\,\left(x^2+y^2+z^2\right)^{3/2}\,\varepsilon_0}+O[d]^3$

The ideal dipole formula in Cartesian Coordinates is

$$V=\frac{qzd}{4\pi\varepsilon_0(x^2+y^2+z^2)^{3/2}}.$$

Example 3.12 Calculate the electric field of the ideal dipole.

In[24]:= $V=\frac{q\,z\,d}{4\,\pi\,\left(x^2+y^2+z^2\right)^{3/2}\,\varepsilon_0}$;

```
E = -Grad[V, {x, y, z}]
```

Out[25]= $\left\{\frac{3\,d\,q\,x\,z}{4\,\pi\,\left(x^2+y^2+z^2\right)^{5/2}\,\varepsilon_0},\ \frac{3\,d\,q\,y\,z}{4\,\pi\,\left(x^2+y^2+z^2\right)^{5/2}\,\varepsilon_0},\ \frac{3\,d\,q\,z^2}{4\,\pi\,\left(x^2+y^2+z^2\right)^{5/2}\,\varepsilon_0}-\frac{d\,q}{4\,\pi\,\left(x^2+y^2+z^2\right)^{3/2}\,\varepsilon_0}\right\}$

Example 3.13 Get the ideal dipole potential in spherical coordinates.

```
In[26]:= Clear[r]; $Assumptions = r > 0;
TransformedField["Cartesian" → "Spherical", V,
  {x, y, z} → {r, θ, ϕ}] // Simplify
```

Out[27]= $\frac{d\,q\,\mathrm{Cos}[\theta]}{4\,\pi\,r^2\,\varepsilon_0}$

Example 3.14 Get the ideal dipole field in spherical coordinates.

```
In[28]:= TransformedField["Cartesian" → "Spherical", E,
  {x, y, z} → {r, θ, ϕ}] // Simplify
```

Out[28]= $\left\{\frac{d\,q\,\mathrm{Cos}[\theta]}{2\,\pi\,r^3\,\varepsilon_0},\ \frac{d\,q\,\mathrm{Sin}[\theta]}{4\,\pi\,r^3\,\varepsilon_0},\ 0\right\}$

3.11 CONDUCTORS

Conductors are materials in which one or more electrons per atom are free to move. The electrons move instantly if one tries to introduce an electric field, until they are no longer pushed.

3.11.1 General Properties

Conductors have a remarkable number of properties:

a) There is zero electric field inside a conductor. If not, the electrons would experience a net force inside the material and be pushed until the field was zero.

b) The charge density inside the conductor is zero. If that were not true, then the electric field would not be zero.

c) If a conductor is charged, all the charge must be on the surface. The charge arranges itself such that the electric field is zero inside the conducting material.

d) Every part of the conductor is at the same potential. Since the electric field is zero inside the material, the potential difference between any 2 points inside the conductor is zero.

e) The electric field just outside the conductor is perpendicular to the surface. The electric potential may only change in the direction that is perpendicular to the surface.

3.11.2 Field Near the Surface

The electric field near a flat conductor with charge density σ may be found using Gauss's law (Fig. 3.5). Take the Gaussian surface to be a pillbox to extend into the conductor and terminate inside. The field is non-zero only on the face outside.

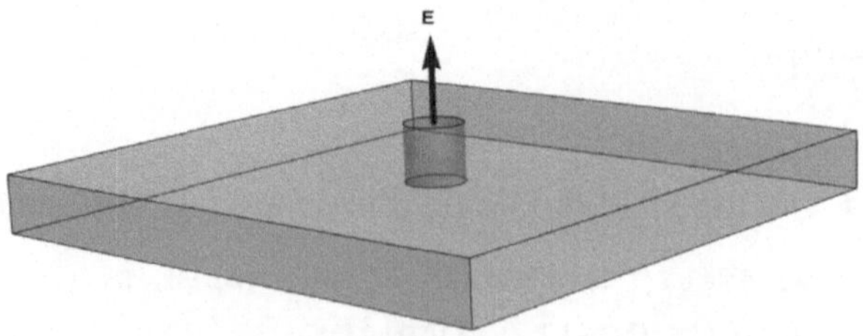

Figure 3.5 The field near a conducting surface appears flat and can be determined using Gauss's law.

$$\int d\mathbf{a} \cdot \mathbf{E} = AE = \frac{\sigma A}{\varepsilon_0}.$$

This gives

$$\mathbf{E} = \frac{\sigma}{\varepsilon_0}\,\hat{\mathbf{n}}.$$

3.12 CAPACITANCE

Capacitance is a measure of the ability of a conductor to hold charge. Consider two conductors with charges Q and $-Q$ that have a potential difference of ΔV. Since the potential difference is proportional to Q, the ratio is constant and is called the capacitance (C),

$$C = \frac{Q}{\Delta V}.$$

The SI unit of capacitance is the farad (F) which is a coulomb per volt.

3.12.1 Capacitance of a Ball or Sphere

Capacitance can also be calculated for a single conductor with respect to another at infinity. The capacitance of a ball or a sphere of radius a is

$$C = \frac{Q}{\left(\frac{Q}{4\pi\varepsilon_0 a}\right)} = 4\pi\varepsilon_0 a.$$

Example 3.15 Calculate the capacitance of the earth.

```
In[29]:= R = Earth PLANET [ average radius ];
         N[UnitConvert[4 π ε₀ R, F], 1]

Out[29]= 0.0007 F
```

One farad is seen to be an enormous capacitance corresponding to an object of size 1000 times the radius of the earth. Capacitance is typically measured in picofarads, corresponding to centimeter distance scales.

Example 3.16 Calculate the capacitance of a 1-cm radius sphere.

```
In[30]:= R = 1 cm; N[UnitConvert[4 π ε₀ R, pF], 3]

Out[30]= 1.11 pF
```

3.12.2 Parallel Plates

Consider a capacitor that is formed with parallel plates of area A separated by distance d. The plates can be any shape but usually are taken to be either disks or squares. The fringe fields can be neglected if $d << \sqrt{A}$. If the plates have charges Q and $-Q$, the magnitude of the electric field between the plates is the superposition of two planes of charge, one positive and the other negative, giving

$$E = \frac{\sigma}{\varepsilon_0} = \frac{Q}{A\varepsilon_0}.$$

The potential difference between the plates is

$$\Delta V = Ed$$

and the capacitance is

$$C = \frac{Q}{\Delta V} = \frac{A\varepsilon_0}{d}.$$

Figure 3.6 Two parallel plates form a capacitor.

Example 3.17 Calculate the capacitance of parallel plate capacitor with $A =$ 1 cm and $d = 1$ mm.

```
In[31]:= A = 1 cm²; d = 1 mm; N[UnitConvert[A ε₀/d, pF], 2]

Out[31]= 0.89 pF
```

Capacitance is significantly enhanced when material is introduced between the plates because the atoms get polarized and produce a contribution to the net electric field that is opposite the original field, thereby reducing ΔV. This is discussed in Chap. 9.

3.12.3 Coax cable

A coax cable consists of a cylindrical wire of radius a with a concentric cylindrical shell of inner radius b (Fig. 3.7).

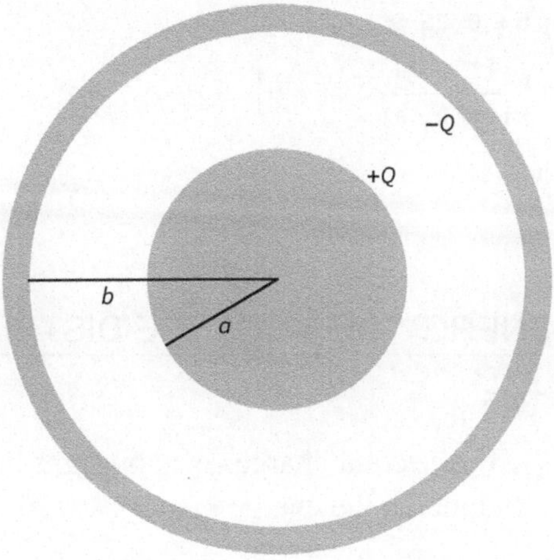

Figure 3.7 A wire and a concentric cylindrical shell (shown here in cross section) form a coax cable.

The potential difference between the inner cylinder and the outer shell is given by the calculation of Ex. 3.4 to be

$$\Delta V = \frac{\lambda \ln\left(\frac{b}{a}\right)}{2\pi\varepsilon_0}.$$

The capacitance for a length L of the coax, using $Q = \lambda L$, is

$$C = \frac{2\pi L\varepsilon_0}{\ln\left(\frac{b}{a}\right)}.$$

This formula has to be modified for the plastic material that separates the conductors (see Sect. 9.3). Writing this additional factor as ε_r, the capacitance per length becomes

$$C/L = \frac{2\pi\varepsilon_0\varepsilon_r}{\ln\left(\frac{b}{a}\right)}.$$

Example 3.18 A coax cable has wire radius of 1.1 mm and and outer conductor inner radius of 0.35 mm. The material between the conductors has $\varepsilon_r = 2$. Calculate the capacitance per length.

```
In[32]:= a = 0.11 mm; b = 0.35 mm; εr = 2;
         UnitConvert[2 π ε0 εr / Log[b / a], pF / m]

Out[33]= 96.1292 pF/m
```

3.13 STORED ENERGY OF A CHARGE DISTRIBUTION

3.13.1 Point Charges

Consider a charge q_1. If a second charge q_2 is brought in from infinity to a distance R_{12}, the work done on the charge is

$$W_2 = \frac{q_1 q_2}{4\pi\varepsilon_0 R_{12}}.$$

If a third charge is brought in from infinity to distance R_{13} from q_1 and R_{23} from q_2, the work done is

$$W_3 = \frac{q_1 q_3}{4\pi\varepsilon_0 R_{13}} + \frac{q_2 q_3}{4\pi\varepsilon_0 R_{23}}.$$

The total stored energy (U) is the work done to assemble the 3 charges,

$$U = W_2 + W_3 = \frac{q_1 q_2}{4\pi\varepsilon_0 R_{12}} + \frac{q_1 q_3}{4\pi\varepsilon_0 R_{13}} + \frac{q_2 q_3}{4\pi\varepsilon_0 R_{23}}.$$

In the assembly of n charges the stored energy is

$$U = \frac{1}{4\pi\varepsilon_1} \sum_{i=0}^{n} \sum_{j>i}^{n} \frac{q_i q_j}{R_{ij}},$$

where R_{ij} is the distance between charges q_i and q_j.

The potential energy of any distribution of charges can be easily calculated by putting the charges and their separations into an arrays and then using the function Sum[f, {i, i_{min}, i_{max}}].

Example 3.19 Calculate the stored energy for the charge configuration in Fig. 3.8 for $q_1 = Q$, $q_2 = -Q/3$, $q_3 = 3Q/5$, and $q_4 = -Q/2$. Take the side of the square to be d.

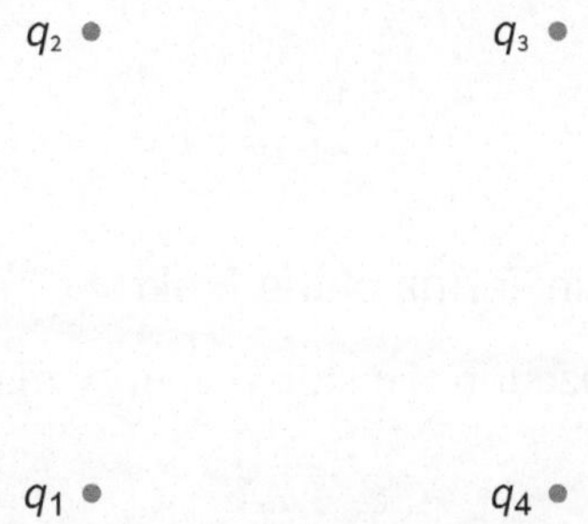

Figure 3.8 Four charges are assembled in a square.

```
In[34]:= ClearAll["Global`*"];
         n = 4; q = {1, -1/3, 3/5, -1/2} Q;
         R = {{0, 1, √2, 1}, {1, 0, 1, √2}, {√2, 1, 0, 1},
            {1, √2, 1, 0}} d;
         1/(4 π ε0) Sum[q[[i]] × q[[j]] / R[[i, j]], {i, 1, n}, {j, i + 1, n}]
```

$$\text{Out[36]=}\quad \frac{-\frac{4\,Q^2}{3\,d} + \frac{23\,Q^2}{30\sqrt{2}\,d}}{4\pi\varepsilon_0}$$

3.13.2 Stored Energy in a Capacitor

Consider the parallel plate capacitor of Fig. 3.6. Suppose that plates are initially uncharged, and negative charge is moved a little bit at a time from the top plate to the bottom plate. The energy dU required to move dq against voltage difference $\Delta V(q)$ which depends on the amount of charge that has been moved is

$$dU = dq\Delta V(q) = dq\frac{q}{C}.$$

The total stored energy after moving charge Q is

$$U = \int_0^Q dq\,\frac{q}{C} = \frac{Q^2}{2C}.$$

Since $C = Q/\Delta V$, the stored energy can also be written

$$U = \frac{1}{2}Q\Delta V$$

or

$$U = \frac{1}{2}C(\Delta V)^2.$$

3.13.3 Stored Energy in Terms of the Field

For the parallel plate capacitor, the stored energy may be written in terms of the field using

$$E = \frac{\Delta V}{d}$$

which gives

$$U = \frac{1}{2}\left(\frac{\varepsilon_0 A}{d}\right)(Ed)^2 = \frac{\varepsilon_0}{2}AdE^2.$$

The stored energy per volume u is

$$u = \frac{U}{Ad} = \frac{\varepsilon_0}{2}E^2.$$

This is an important result that is universally true.

Example 3.20 Calculate the stored energy in the capacitor of Ex. 3.17 if the voltage is 10 V.

```
In[37]:= ΔV = 10 V; A = 1 cm²; d = 1 mm; E = ΔV/d;
         N[UnitConvert[A d ε0/2 E², J], 2]
Out[38]= 4.4×10⁻¹¹ J
```

Example 3.21 How many electrons have been moved to make 10 V on the capacitor?

```
In[39]:= N[UnitConvert[A ε0/d ΔV, e], 2]
Out[39]= 5.5×10⁷ e
```

CHAPTER 4

The Biot-Savart Law

4.1 MAGNETIC FORCE AND MAGNETIC FIELD

No magnetic charge has ever been observed. Magnetic forces occur when electric charges move. Consider the two moving charges of Fig. 4.1. Both charges have to be moving to have be a magnetic force.

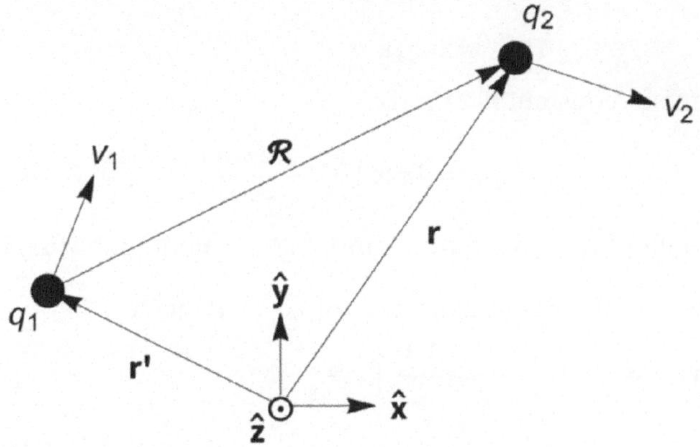

Figure 4.1 Two moving charges experience magnetic forces that depend on a double cross-product of the velocity vectors and the separation vector $\mathcal{R}$.

Analogous to the electric case, the concept of magnetic field is used to describe this force. Most of this chapter is concerned with how to calculate magnetic fields from moving charges in the form of a steady current. Once the field is known, the magnetic force on an additional charge q moving with velocity vector $\mathbf{v}$ placed in a magnetic field $\mathbf{B}$ is

$$\mathbf{F}_{\mathrm{m}} = q\mathbf{v} \times \mathbf{B},$$

and the total force, electric plus magnetic is

$$\mathbf{F} = q(\mathbf{E} + \mathbf{v} \times \mathbf{B}).$$

This is known as the Lorentz force. Note that the charge q is not part of the determination of the fields.

If the charges are not relativistic, then the field caused by q_1 is

$$\mathbf{B} = \frac{\mu_0 q_1 \mathbf{v}_1 \times \boldsymbol{\mathcal{R}}}{4\pi},$$

where $\boldsymbol{\mathcal{R}}$ is the vector that point from the location of q_1 to the place where the field is evaluated.

Example 4.1 Get the magnetic constant.

```
In[1]:= Quantity["MagneticConstant"]

Out[1]= μ0
```

Example 4.2 Get the numerical value of the magnetic constant.

```
In[2]:= UnitConvert[μ0]

Out[2]= 1.256637062 × 10^-6 kg m/(s^2 A^2)
```

The magnetic constant is exactly

$$\mu_0 = 4\pi \times 10^{-7}\,\frac{\mathrm{kg\cdot m}}{\mathrm{s^2\cdot A}},$$

from the SI definition of the amp. A more convenient unit of μ_0 is $\mathrm{T\cdot m/A}$.

Example 4.3 Get the numerical value of $\frac{\mu_0}{4\pi}$ in $\mathrm{T\cdot m/A}$.

```
In[3]:= N[UnitConvert[μ0/(4 π), (T m)/A], 9]

Out[3]= 1.00000000 × 10^-7 m T/A
```

Example 4.4 An elementary charge at the origin moves with speed 10^6 m/s in the z direction. Calculate the magnetic field along the x-axis at a distance of 1 nm.

```
In[4]:= q = e; v = {0, 0, 1} 10^6 m/s; R = {1, 0, 0} nm;

        B = N[UnitConvert[(μ0 q v × R)/(4 π (R.R)^(3/2)), T], 3]

Out[5]= {0 T, 0.0160 T, 0 T}
```

The force $\mathbf{F}_2$ on q_2 is given by the Lorentz force, where $\mathbf{B}$ is the field caused by q_1 at the location of q_2,

$$\mathbf{F}_2 = \left(\frac{\mu_0 q_1 q_2}{4\pi \mathcal{R}^3}\right)\mathbf{v}_2 \times (\mathbf{v}_1 \times \boldsymbol{\mathcal{R}}).$$

This is an experimental result like Coulomb's law. Both Coulomb's law and this magnetic force law are not valid relativistically.

Example 4.5 At the location of the magnetic field **B** calculated in ex. 4.4, another elementary charge Q has a velocity vector $\mathbf{V} = (1,2,3) \times 10^6$ m/s. Calculate the magnetic force on Q.

```
In[6]:= Q = e; V = {1, 2, 3} 10^6 m/s;

        N[UnitConvert[Q V × B, N], 3]

Out[7]= {-7.70×10^-15 N, 0 N, 2.57×10^-15 N}
```

The magnetic force of q_2 on q_1 is

$$\mathbf{F}_1 = \left(\frac{\mu_0 q_1 q_2}{4\pi \mathcal{R}^3}\right)\mathbf{v}_1 \times [\mathbf{v}_2 \times (-\boldsymbol{\mathcal{R}})],$$

The magnetic force $\mathbf{F}_2$ does not equal to $-\mathbf{F}_1$ as might be expected from Newton's third law. The third law does not generally hold in special relativity because the forces on q_1 and q_2 do not even occur at the same time. The expressions for the forces are only a non-relativistic approximation.

4.2 MAGNETIC FLUX

Even though there is no magnetic charge, the concept of magnetic flux is equally useful to that of electric flux and is defined by

$$\Phi = \int da\hat{\mathbf{n}} \cdot \mathbf{B}.$$

As in the electric case, there are two choices for the unit vector normal to the surface and the choice gives the sign of the flux. For a closed surface, $\hat{\mathbf{n}}$ is chosen to be outward.

4.3 MAXWELL EQUATION

Gauss's law for magnetism reads

$$\oint d\mathbf{a} \cdot \mathbf{B} = 0.$$

This is one of the four fundamental Maxwell equations and is often referred to as Gauss's law for magnetic fields. It is relativistically correct.

The differential form is deceptively simple,

$$\nabla \cdot \mathbf{B} = 0.$$

The physics of Gauss's law for magnetic fields is that since there is no source of the field analogous to electric charge, the magnetic field can never diverge. This means that magnetic field lines must be closed loops. They have no beginning and no end.

4.4 ELECTRIC CURRENT AND THE BIOT-SAVART LAW

Moving charge is electric current and is measured in C/s which is called an amp (A). A steady current consists of a group of charges moving together such that the charge density remains constant. Moving point charges are obviously not a steady current. The simplest example of a steady current is a closed loop of moving charge.

The expression $q\mathbf{v}$ is a current I times a displacement vector $d\boldsymbol{\ell}$ which follows the path of the moving charge. This is referred to as a current element. The current can be written as a vector and

$$\mathbf{I}d\ell = Id\boldsymbol{\ell}.$$

It does not matter if the vector goes with the current or the displacement because they are in the same direction. The Biot-Savart law states that the contribution to the magnetic field from a current element is

$$d\mathbf{B} = \frac{\mu_0 (Id\boldsymbol{\ell}) \times \boldsymbol{\mathcal{R}}}{4\pi \mathcal{R}^3}.$$

This gives the field for a non-relativistic moving point charge with $Id\boldsymbol{\ell} = q\mathbf{v}$. The total field for an extended current is obtained by integration,

$$\mathbf{B} = \frac{\mu_0}{4\pi} \int \frac{(Id\boldsymbol{\ell}) \times \boldsymbol{\mathcal{R}}}{\mathcal{R}^3}.$$

This is the experimentally determined magnetic parallel of the electric field defined by Coulomb's law. It is even relativistically correct as long as the current is steady.

Since the Maxwell equations are relativistically correct, it is perhaps not too surprising that the speed of light c is somehow present. It is given by

$$c = \frac{1}{\sqrt{\epsilon_0 \mu_0}}.$$

The speed of light is obtained with Quantity[“SpeedOfLight”].

Example 4.6 Get the speed of light.

```
In[8]:= Quantity["SpeedOfLight"]

Out[8]= c
```

Example 4.7 Get the numerical value of the speed of light.

```
In[9]:= UnitConvert[c]

Out[9]= 299 792 458 m/s
```

The “==” does a logical comparison.

Example 4.8 Compare the speed of light to $1/\sqrt{\epsilon_0\mu_0}$.

In[10]:= $c == \dfrac{1}{\sqrt{\varepsilon_0\,\mu_0}}$

Out[10]= True

4.5 CONTINUITY EQUATION

The continuity equation is a relationship between charge and current and is an expression of local charge conservation. In words, it says that charge can only be decreasing (increasing) in a region if there is a current that carries the charge out of (into) the region. Mathematically,

$$\nabla \cdot \mathbf{J} = -\frac{\partial \rho}{\partial t}.$$

Taking the volume integral of both sides and using the divergence theorem,

$$\int dv'\, \nabla \cdot \mathbf{J} = \oint d\mathbf{a}' \cdot \mathbf{J} = -\frac{\partial q}{\partial t},$$

where q is the volume integral of the charge density. The flux of current density through a closed surface that surrounds a volume of charge is equal to the rate that charge leaves the volume. The continuity equation is contained in the complete Maxwell equations, and it motivates the need for a term with the time derivative of the electric field in Ampère’s law (Sect. 10.1).

4.6 LINE OF CURRENT

The current vector **I** may be written in terms of the linear charge density λ and velocity **v**,

$$\mathbf{I} = \lambda \mathbf{v}.$$

Consider a segment of a wire, which is a part of steady current, that extends from $-L/2$ to $L/2$ along the z-axis (Fig. 4.2).

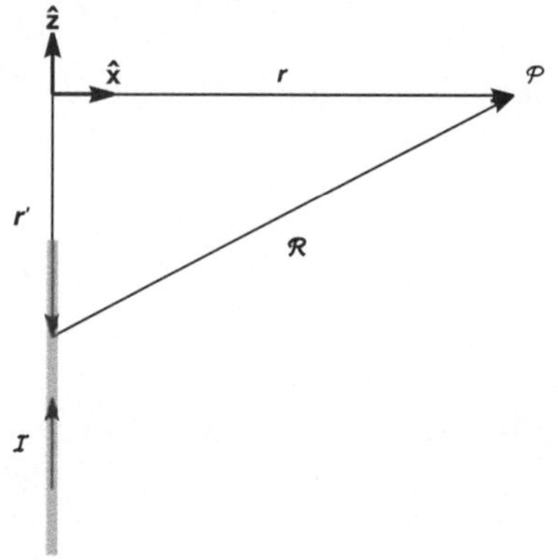

Figure 4.2 A segment of current extends along the z-axis.

Example 4.9 Calculate the magnetic field as a function of (x, y) at $z = 0$.

```
In[11]:= $Assumptions = {L > 0, z′ ∈ ℝ, z > 0, x > 0, y > 0};
         r = {x, y, 0}; r′ = {0, 0, z′}; ℛ = r - r′;
```

$$B = \frac{\mu_0}{4\pi} \int_{-L/2}^{L/2} \frac{I \{0, 0, 1\} \times \mathcal{R}}{(\mathcal{R}.\mathcal{R})^{3/2}} \, dz'$$

Out[13]= $\left\{-\frac{L\, y\, I\, \mu_0}{2\pi (x^2 + y^2)\sqrt{L^2 + 4(x^2 + y^2)}}, \frac{L\, x\, I\, \mu_0}{2\pi (x^2 + y^2)\sqrt{L^2 + 4(x^2 + y^2)}}, 0\right\}$

Example 4.10 Transform, the field into cylindrical coordinates.

```
In[14]:= Clear[r];
         f = TransformedField["Cartesian" → "Cylindrical",
           B, {x, y, z} → {r, ϕ, z′}] // Simplify
```

Out[14]= $\left\{0, \frac{L\, I\, \mu_0}{2\pi r \sqrt{L^2 + 4 r^2}}, 0\right\}$

Example 4.11 Take the limit as $L \to \infty$.

```
In[15]:= Limit[f, L → ∞]
```

$$\text{Out[15]=}\ \left\{0,\ \frac{I\,\mu_0}{2\,\pi\,r},\ 0\right\}$$

Example 4.12 Calculate the magnetic field as a function of (x, y, z).

```
In[16]:= $Assumptions = {L > 0, z′ ∈ ℝ, z > 0, x > 0, y > 0};
         r = {x, y, z}; r′ = {0, 0, z′}; ℛ = r - r′;
```

$$B = \frac{\mu_0}{4\,\pi}\int_{-L/2}^{L/2}\frac{I\,\{0,\,0,\,1\}\times\mathcal{R}}{(\mathcal{R}.\mathcal{R})^{3/2}}\,dz'$$

$$\text{Out[18]=}\ \left\{-\frac{1}{4\,\pi\,(x^2+y^2)}\,y\left(2\,z\left(-\frac{1}{\sqrt{4\,(x^2+y^2)+(L-2\,z)^2}}+\frac{1}{\sqrt{4\,(x^2+y^2)+(L+2\,z)^2}}\right)+L\left(\frac{1}{\sqrt{4\,(x^2+y^2)+(L-2\,z)^2}}+\frac{1}{\sqrt{4\,(x^2+y^2)+(L+2\,z)^2}}\right)\right)I\,\mu_0,\right.$$

$$\frac{1}{4\,\pi\,(x^2+y^2)}\,x\left(2\,z\left(-\frac{1}{\sqrt{4\,(x^2+y^2)+(L-2\,z)^2}}+\frac{1}{\sqrt{4\,(x^2+y^2)+(L+2\,z)^2}}\right)+L\left(\frac{1}{\sqrt{4\,(x^2+y^2)+(L-2\,z)^2}}+\frac{1}{\sqrt{4\,(x^2+y^2)+(L+2\,z)^2}}\right)\right)$$

$$\left.I\,\mu_0,\ 0\right\}$$

Example 4.13 Take the limit as $L \to \infty$.

```
In[19]:= f = Limit[B, L → ∞]
```

$$\text{Out[19]=}\ \left\{-\frac{y\,I\,\mu_0}{2\,\pi\,(x^2+y^2)},\ \frac{x\,I\,\mu_0}{2\,\pi\,x^2+2\,\pi\,y^2},\ 0\right\}$$

Example 4.14 Transform, the field into cylindrical coordinates.

```
In[20]:= Clear[r];
         TransformedField["Cartesian" → "Cylindrical", f,
           {x, y, z} → {r, ϕ, z'}] // Simplify
```

$$\text{Out[20]}= \left\{0, \frac{I\mu_0}{2\pi r}, 0\right\}$$

4.7 CYLINDRICAL SHELL OF CURRENT

Consider a current in the z-direction in the shape of a cylindrical shell (Fig. 4.3). Such a 2-dimensional current can be written in terms of a surface current density vector **K** which is the current per distance perpendicular to the current. In this case,

$$\mathbf{K} = \frac{\mathbf{I}}{2\pi a},$$

where a is the radius of the cylinder. The Biot-Savart law reads

$$\mathbf{B} = \frac{\mu_0}{4\pi} \int dA \frac{\mathbf{K} \times \boldsymbol{\mathcal{R}}}{\mathcal{R}^3}.$$

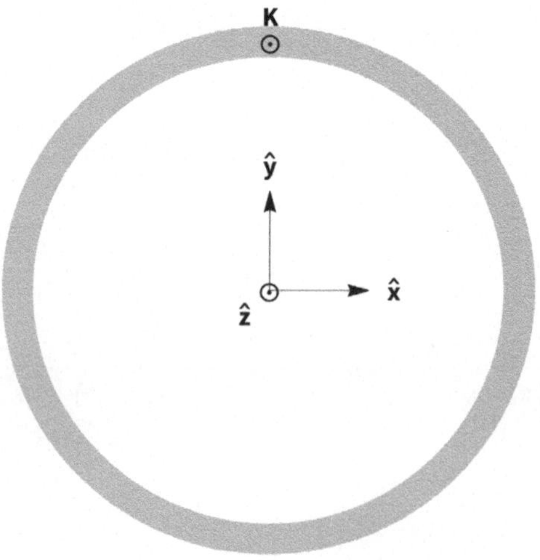

Figure 4.3 A surface current **K** is along a cylindrical shell..

Example 4.15 Calculate the magnetic field outside a cylindrical shell of current. The direction of the current is along the axis of the cylinder.

```
In[21]:= $Assumptions = x > a > 0;
         r = {x, 0, 0}; r′ = {a Cos[ϕ′], a Sin[ϕ′], z′};
         ℛ = r - r′;
```

$$f = \frac{\mu_0}{4\pi}\int \frac{\frac{\mathcal{I}}{2\pi a}\,\{0,\,0,\,1\}\times\mathcal{R}}{(\mathcal{R}.\mathcal{R})^{3/2}}\,dz' \text{ // Simplify;}$$

```
         g = Integrate[(Limit[f, z′ → ∞] - Limit[f, z′ → -∞]) a,
           {ϕ′, 0, 2 π}]
```

$$\text{Out[23]= } \left\{0,\ \frac{\mathcal{I}\,\mu_0}{2\,\pi\,x},\ 0\right\}$$

Example 4.16 Calculate the magnetic field inside a cylindrical shell of current.

```
In[24]:= $Assumptions = {x < a, a > 0, x > 0};
         r = {x, 0, 0}; r′ = {a Cos[ϕ′], a Sin[ϕ′], z′};
         ℛ = r - r′;
```

$$f = \frac{\mu_0}{4\pi}\int \frac{\frac{\mathcal{I}}{2\pi a}\,\{0,\,0,\,1\}\times\mathcal{R}}{(\mathcal{R}.\mathcal{R})^{3/2}}\,dz' \text{ // Simplify;}$$

```
         g = Integrate[(Limit[f, z′ → ∞] - Limit[f, z′ → -∞]) a,
           {ϕ′, 0, 2 π}]

Out[26]= {0, 0, 0}
```

4.8 CURRENT LOOP

Consider a circular current loop of radius a in the $x-y$ plane (Fig. 4.4). To integrate around a loop, use Cartesian coordinates and the polar angle variable of cylindrical coordinates (Ex. C.8),

$$\hat{\boldsymbol{\phi}} = -\sin\phi\,\hat{\mathbf{x}} + \cos\phi\,\hat{\mathbf{y}}$$

to get

$$Id\boldsymbol{\ell} = Iad\boldsymbol{\phi} = Ia(-\sin\phi\,\hat{\mathbf{x}} + \cos\phi\,\hat{\mathbf{y}})$$

At an arbitrary location, the exact field from the loop is complicated, involving elliptic integrals which can occur when having the square root of a polynomial function in the integrand.

Example 4.17 Find the magnetic field at an arbitrary distance from a current loop.

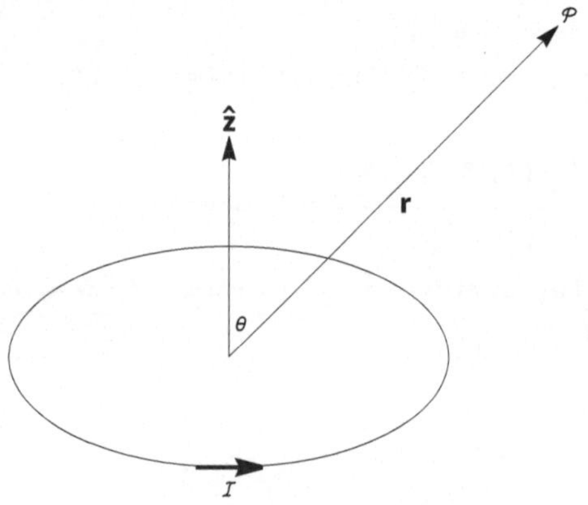

Figure 4.4 A circular current loop lies in the $x-y$ plane.

```
In[27]:= ClearAll["Global`*"];
         r = {x, 0, z}; r′ = a {Cos[ϕ′], Sin[ϕ′], 0};
         ℛ = r - r′;
         f[ϕ′_] = μ0 I/(4 π) ∫ ({-Sin[ϕ′], Cos[ϕ′], 0} × ℛ)/(ℛ.ℛ)^(3/2) a dϕ′ //
            Simplify;
         B = f[2 π] - f[0]
```

$$\text{Out[29]= } \left\{\left(z\,I\left(2\left(a^2+x^2+z^2\right)\text{EllipticE}\left[-\frac{4\,a\,x}{a^2-2\,a\,x+x^2+z^2}\right]-2\left(a^2+2\,a\,x+x^2+z^2\right)\text{EllipticK}\left[-\frac{4\,a\,x}{a^2-2\,a\,x+x^2+z^2}\right]\right)\mu_0\right)\Big/\left(4\,\pi\,x\,\sqrt{a^2-2\,a\,x+x^2+z^2}\,\left(a^2+2\,a\,x+x^2+z^2\right)\right),\right.$$

$$\left. 0,\ \left(I\left(2\left(a^2-x^2-z^2\right)\text{EllipticE}\left[-\frac{4\,a\,x}{a^2-2\,a\,x+x^2+z^2}\right]+2\left(a^2+2\,a\,x+x^2+z^2\right)\text{EllipticK}\left[-\frac{4\,a\,x}{a^2-2\,a\,x+x^2+z^2}\right]\right)\mu_0\right)\Big/\left(4\,\pi\,\sqrt{a^2-2\,a\,x+x^2+z^2}\,\left(a^2+2\,a\,x+x^2+z^2\right)\right)\right\}$$

The integral may always be evaluated numerically.

Example 4.18 Find the magnetic field at $(x, y, z) = (4,0,6)$ cm from a 2-cm radius loop carrying 1 A.

```
In[30]:= x = 04; z = 0.06; a = 0.02; I = 1 A;
         r = {x, 0, z}; r′ = a { Cos[ϕ′], Sin[ϕ′], 0 };
         ℛ = r - r′;
         UnitConvert[
           μ0 I                 {-Sin[ϕ′], Cos[ϕ′], 0 } × ℛ
          ────── NIntegrate[ ───────────────────────────── a ,
           4 π m                      (ℛ.ℛ)^(3/2)
            {ϕ′, 0, 2 π}], T]

Out[32]= {8.83117 × 10^-14 T, -1.58819 × 10^-28 T, -1.96156 × 10^-12 T}
```

Along the axis of the loop, the field is simple.

Example 4.19 Calculate the magnetic field along the axis of a current loop.

```
In[33]:= ClearAll["Global`*"];
         r = {0, 0, z}; r′ = a {Cos[ϕ′], Sin[ϕ′], 0};
         ℛ = r - r′;
                  μ0 I  ⌠ {-Sin[ϕ′], Cos[ϕ′], 0} × ℛ
         f[ϕ′_] = ──── │ ─────────────────────────── a ⅆϕ′ //
                  4 π   ⌡        (ℛ.ℛ)^(3/2)
           Simplify;
         B = f[2 π] - f[0]
```

$$\text{Out[35]= } \left\{0,\ 0,\ \frac{a^2\, I\, \mu_0}{2\left(a^2+z^2\right)^{3/2}}\right\}$$

Example 4.20 Calculate the magnetic field at the center of a current loop.

```
In[36]:= B /. z → 0 // Simplify
```

$$\text{Out[36]= } \left\{0,\ 0,\ \frac{I\, \mu_0}{2\, a}\right\}$$

4.9 SOLENOID

The field along the axis of a solenoid may be calculated by adding rings using the result of Ex. 4.19. (The field everywhere is evaluated in Sect. 5.3.4 using Ampère's law.) A differential portion of the field is given by

$$dB = nIdz,$$

where n is the number of loops per unit length along the axis.

Example 4.21 Calculate the magnetic field along the axis and at the center of a solenoid that extends from $-L/2$ to $L/2$.

```
In[37]:= $Assumptions = {L > 0, a > 0}; B = ∫_{-L/2}^{L/2} (a^2 n I μ0)/(2 (a^2 + z'^2)^(3/2)) dz'

Out[37]= (L n I μ0)/Sqrt[4 a^2 + L^2]
```

Example 4.22 Take the limit for an infinitely long solenoid.

```
In[38]:= Series[B, {L, ∞, 1}]

Out[38]= n I μ0 + O[1/L]^2
```

A second way to view the solenoid is that of a cylindrical surface current in the $\hat{\boldsymbol{\phi}}$ direction and directly use the Biot-Savart law to integrate over the surface of the cylinder. This is easiest in cylindrical coordinates.

Example 4.23 Repeat the calculation of the finite solenoid of Ex. 4.21 with direct integration.

```
In[39]:= r = {0, 0, 0}; r' = {a Cos[ϕ'], a Sin[ϕ'], z'};
         ℛ = r - r';
         μ0 (I / L)/(4 π) ∫_{-L/2}^{L/2} (∫_0^{2π} ({-Sin[ϕ'], Cos[ϕ'], 0} × ℛ)/(ℛ.ℛ)^(3/2) a dϕ') dz' //
           Simplify

Out[40]= {0, 0, (I μ0)/Sqrt[4 a^2 + L^2]}
```

In this calculation I is the total current on the surface. To compare to Ex. 4.21, $I \rightarrow nLI$.

The solenoid can be integrated numerically for an arbitrary position both inside and outside. Without loss of generality, one can take $y = 0$, $a = 1$, and chose a random position for x.

Example 4.24 Calculate the field inside an infinite solenoid at a random position.

```
In[41]:= r = {Random[], 0, 0}; r′ = { Cos[ϕ′], Sin[ϕ′], z′};
         ℛ = r - r′;
         μ0 (I / L)              {-Sin[ϕ′], Cos[ϕ′], 0} × ℛ
         ────────── NIntegrate[ ─────────────────────────── ,
            4 π                         (ℛ.ℛ)^(3/2)
           {ϕ′, 0, 2 π}, {z′, -∞, ∞}] /. I → n I L

Out[42]= {0., 0., 1. n I μ0}
```

Example 4.25 Calculate the field outside an infinite solenoid at a random position.

```
In[43]:= r = {1 + 100 Random[], 0, 0}; r′ = { Cos[ϕ′], Sin[ϕ′], z′};
         ℛ = r - r′;
         μ0 (I / L)              {-Sin[ϕ′], Cos[ϕ′], 0} × ℛ
         ────────── NIntegrate[ ─────────────────────────── ,
            4 π                         (ℛ.ℛ)^(3/2)
           {ϕ′, 0, 2 π}, {z′, -∞, ∞}] /. I → n I L

Out[44]= {0., 0., -1.36849 × 10^-19 n I μ0}
```

This is a remarkable result. The field is uniform everywhere inside the solenoid and zero everywhere outside the solenoid.

4.10 MAGNETIC DIPOLE

The square loop can be used to arrive at the formula for an ideal magnetic dipole, where the observation distance is much larger than the dimension of the current loop. The calculation of the square loop is just the sum of four straight wire segments as calculated in Ex. 4.12.

Example 4.26 Calculate the magnetic field due to a square current loop.

```
In[45]:= ClearAll["Global`*"]; $Assumptions = {L > 0};
        r = {x, y, z};
        BLine[r′_, dir_] :=
         With[{ℛ = r - r′, xy = r′[[dir]]},
          ((# /. xy → L/2) - (# /. xy → -L/2)) &[
            μ0   ⌠ I UnitVector[3, dir] × ℛ
            ── ⌡ ───────────────────────── d(xy)]];
            4 π        (ℛ.ℛ)^(3/2)
        B = BLine[{L / 2, y′, 0}, 2] - BLine[{x′, L / 2, 0}, 1] -
           BLine[{-L / 2, y′, 0}, 2] + BLine[{x′, -L / 2, 0}, 1] //
          FullSimplify
```

Out[47]=

$$\Bigg\{\frac{1}{\pi}\sqrt{2}\,z\left(\frac{-L+2y}{2\left((L+2x)^2+4z^2\right)\sqrt{L^2+2L(x-y)+2\left(x^2+y^2+z^2\right)}}+\frac{L+2y}{2\left((L-2x)^2+4z^2\right)\sqrt{L^2+2L(-x+y)+2\left(x^2+y^2+z^2\right)}}+\frac{L-2y}{2\left((L-2x)^2+4z^2\right)\sqrt{L^2-2L(x+y)+2\left(x^2+y^2+z^2\right)}}+\frac{-L-2y}{2\left((L+2x)^2+4z^2\right)\sqrt{L^2+2L(x+y)+2\left(x^2+y^2+z^2\right)}}\right)\mathcal{I}\,\mu_0,$$

$$\frac{1}{\pi}\sqrt{2}\,z\left(\frac{L+2x}{2\left((L-2y)^2+4z^2\right)\sqrt{L^2+2L(x-y)+2\left(x^2+y^2+z^2\right)}}+\frac{-L+2x}{2\left((L+2y)^2+4z^2\right)\sqrt{L^2+2L(-x+y)+2\left(x^2+y^2+z^2\right)}}+\frac{L-2x}{2\left((L-2y)^2+4z^2\right)\sqrt{L^2-2L(x+y)+2\left(x^2+y^2+z^2\right)}}+\frac{-L-2x}{2\left((L+2y)^2+4z^2\right)\sqrt{L^2+2L(x+y)+2\left(x^2+y^2+z^2\right)}}\right)\mathcal{I}\,\mu_0,$$

$$\frac{1}{\sqrt{2}\,\pi}\left(\frac{(L+2x)(L-2y)}{2\left((L+2x)^2+4z^2\right)\sqrt{L^2+2L(x-y)+2\left(x^2+y^2+z^2\right)}}+\frac{(L+2x)(L-2y)}{2\left((L-2y)^2+4z^2\right)\sqrt{L^2+2L(x-y)+2\left(x^2+y^2+z^2\right)}}+\frac{(L-2x)(L+2y)}{2\left((L-2x)^2+4z^2\right)\sqrt{L^2+2L(-x+y)+2\left(x^2+y^2+z^2\right)}}+\frac{(L-2x)(L+2y)}{2\left((L+2y)^2+4z^2\right)\sqrt{L^2+2L(-x+y)+2\left(x^2+y^2+z^2\right)}}+\frac{(L-2x)(L-2y)}{2\left((L-2x)^2+4z^2\right)\sqrt{L^2-2L(x+y)+2\left(x^2+y^2+z^2\right)}}+\frac{(L-2x)(L-2y)}{2\left((L-2y)^2+4z^2\right)\sqrt{L^2-2L(x+y)+2\left(x^2+y^2+z^2\right)}}+\frac{(L+2x)(L+2y)}{2\left((L+2x)^2+4z^2\right)\sqrt{L^2+2L(x+y)+2\left(x^2+y^2+z^2\right)}}+\frac{(L+2x)(L+2y)}{2\left((L+2y)^2+4z^2\right)\sqrt{L^2+2L(x+y)+2\left(x^2+y^2+z^2\right)}}\right)\mathcal{I}\,\mu_0\Bigg\}$$

The output of Ex. 4.26 is exact. (Try doing that without Mathematica!). One can make sense of it by taking the limit of small L. It is not good enough to take the limit as $L \to 0$ because that just gives zero. A series expansion is needed to extract the leading term.

Example 4.27 Find the leading non-zero term for small L.

In[48]:=
```
Series[B, {L, 0, 2}]
```

Out[48]=
$$\left\{\frac{3\,x\,z\,\mathcal{I}\,\mu_0\,L^2}{4\,\pi\,(x^2+y^2+z^2)^{5/2}}+O[L]^3,\ \frac{3\,y\,z\,\mathcal{I}\,\mu_0\,L^2}{4\,\pi\,(x^2+y^2+z^2)^{5/2}}+O[L]^3,\ -\frac{\left((x^2+y^2-2\,z^2)\,\mathcal{I}\,\mu_0\right)L^2}{4\,\left(\pi\,(x^2+y^2+z^2)^{5/2}\right)}+O[L]^3\right\}$$

This is the pure dipole formula, sometimes referred to as the "ideal" or "perfect" dipole in Cartesian coordinates with magnetic dipole moment **m** equal to

$$\mathbf{m} = IL^2\hat{\mathbf{z}}.$$

It is perhaps more easily recognized in spherical coordinates.

Example 4.28 Transform to spherical coordinates.

In[49]:=
```
ClearAll["Global`*"];
$Assumptions = r > 0;
```

$$B = \left\{\frac{3\,x\,z\,\mathcal{I}\,\mu_0\,L^2}{4\,\pi\,(x^2+y^2+z^2)^{5/2}},\ \frac{3\,y\,z\,\mathcal{I}\,\mu_0\,L^2}{4\,\pi\,(x^2+y^2+z^2)^{5/2}},\ -\frac{(x^2+y^2-2\,z^2)\,\mathcal{I}\,\mu_0\,L^2}{4\,\pi\,(x^2+y^2+z^2)^{5/2}}\right\};$$

```
TransformedField["Cartesian" → "Spherical", B,
  {x, y, z} → {r, θ, ϕ}] // Simplify
```

Out[50]=
$$\left\{\frac{L^2\,\mathcal{I}\,\mathrm{Cos}[\theta]\,\mu_0}{2\,\pi\,r^3},\ \frac{L^2\,\mathcal{I}\,\mathrm{Sin}[\theta]\,\mu_0}{4\,\pi\,r^3},\ 0\right\}$$

The field for the magnetic dipole field is derived again in a completely different way in Sect. 6.2.3 using the vector potential.

The magnetic dipole field has the same form as the electric dipole field (Ex. 3.14) with

$$\frac{1}{\varepsilon_0} \to \mu_0,$$

and the dipole strength

$$qd \rightarrow IL^2.$$

In each case the direction of the dipole vector is along the z direction, which defines the polar angle θ (App. C).

A straightforward (if somewhat lengthy) calculation shows that the magnetic field can be put in the so-called coordinate independent form using

$$\hat{\mathbf{r}} = \sin\theta\cos\phi\,\hat{\mathbf{x}} + \sin\theta\sin\phi\,\hat{\mathbf{y}} + \cos\theta\,\hat{\mathbf{z}}$$

and

$$\hat{\mathbf{z}} = \cos\theta\,\hat{\mathbf{r}} - \sin\theta\,\hat{\boldsymbol{\theta}}.$$

which gives

$$\mathbf{B} = \frac{\mu_0}{4\pi}\left[\frac{3(\mathbf{m}\cdot\hat{\mathbf{r}})\hat{\mathbf{r}} - \mathbf{m}}{r^3}\right],$$

where $\mathbf{m}$ is taken to be in the z direction.

In Ex. 6.21 it is shown that the coordinate independent form of the dipole field gives the same result as Ex. 4.28.

The electric version is

$$\mathbf{E} = \frac{1}{4\pi\varepsilon_0}\left[\frac{3(\mathbf{p}\cdot\hat{\mathbf{r}})\hat{\mathbf{r}} - \mathbf{p}}{r^3}\right],$$

where $\mathbf{p}$ is taken to be in the z direction.

CHAPTER 5

Ampere's Law

Ampère's law states that the line integral of **B** around a closed loop is equal to the magnetic constant times the current enclosed by the loop,

$$\oint d\boldsymbol{\ell}\cdot\mathbf{B} = \mu_0 I_{\mathrm{en}}.$$

Ampère's law may be derived from the Biot-Savart law as is done in Sect. 5.2. In 5.1, the validity of Ampère's law is verified with a variety of examples.

The simplest case is a long straight wire with a steady current. Outside the wire, the magnetic field was calculated in cylindrical coordinates (Ex. 4.14) to be

$$\mathbf{B} = \frac{\mu_0 I}{2\pi r}\hat{\boldsymbol{\phi}}.$$

The curl is zero.

Example 5.1 Calculate the curl of $\hat{\boldsymbol{\phi}}/r$.

```
In[1]:= Curl[{0, 1/r, 0}, {r, ϕ, z}, "Cylindrical"]

Out[1]= {0, 0, 0}
```

This is analogous to the divergence of the field due to a point charge which is zero everywhere except at the location of the charge where it becomes infinite. In this case, the curl becomes infinite at the location of the current (inside the wire), giving

$$\boldsymbol{\nabla}\times\mathbf{B} = \mu_0\mathbf{J},$$

which is Ampère's law in differential form. (Compare to $\boldsymbol{\nabla}\cdot\mathbf{E} = \rho/\varepsilon_0$.) Integrating over an arbitrary surface gives

$$\int dA\,\hat{\mathbf{n}}\cdot(\boldsymbol{\nabla}\times\mathbf{B}) = \mu_0\int dA\,\hat{\mathbf{n}}\cdot\mathbf{J}.$$

Using the mathematical identity called Stokes' theorem,

$$\int dA\, \hat{\mathbf{n}} \cdot (\boldsymbol{\nabla} \times \mathbf{B}) = \oint d\boldsymbol{\ell} \cdot \mathbf{B},$$

where the area in the integral on the left is enclosed by the line integral on the right. This gives Ampère's law in integral from,

$$\oint d\boldsymbol{\ell} \cdot \mathbf{B} = \mu_0 \int dA\, \hat{\mathbf{n}} \cdot \mathbf{J} = \mu_0 I_{\text{en}}.$$

The physics statement of Ampère's law is that the line integral of the magnetic field around any closed loop is equal to μ_0 times the current that passes through the loop.

5.1 EXAMPLES OF AMPÈRE'S LAW

5.1.1 Square Loop

In Ex. 4.9 the field was calculated at the midpoint of a current segment,

$$B = -\frac{\mu_0 I y}{2\pi(x^2 + y^2)}\hat{\mathbf{x}} + \frac{\mu_0 I x}{2\pi(x^2 + y^2)}\hat{\mathbf{y}}.$$

Now consider the field at the center of a square loop (5.1).

Example 5.2 Integrate the **B** field around a square loop of arbitrary side ($2a$) for a long wire at its center.

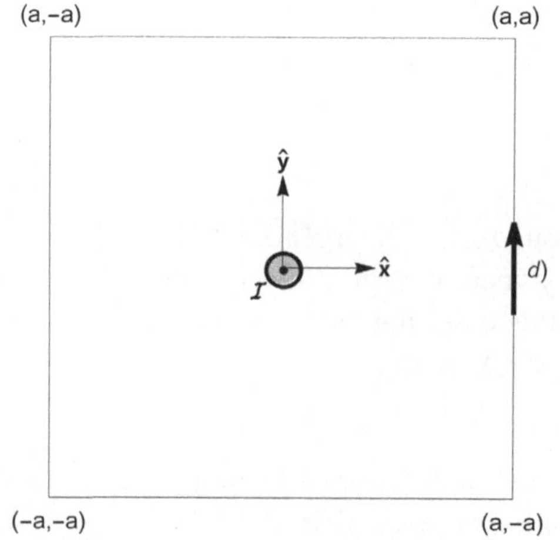

Figure 5.1 The integration path for Ex. 5.2 is a square loop of arbitrary size that encloses the current.

```
In[2]:= B = {-y I μ0/(2 π (x^2 + y^2)), x I μ0/(2 π (x^2 + y^2)), 0};
        ∫_{-a}^{a} (B[[2]] /. x → a) ⅆy + ∫_{a}^{-a} (B[[1]] /. y → a) ⅆx +
        ∫_{a}^{-a} (B[[2]] /. x → -a) ⅆy + ∫_{-a}^{a} (B[[1]] /. y → -a) ⅆx

Out[3]= I μ0
```

Figure 5.2 shows an integration path that does not enclose the current.

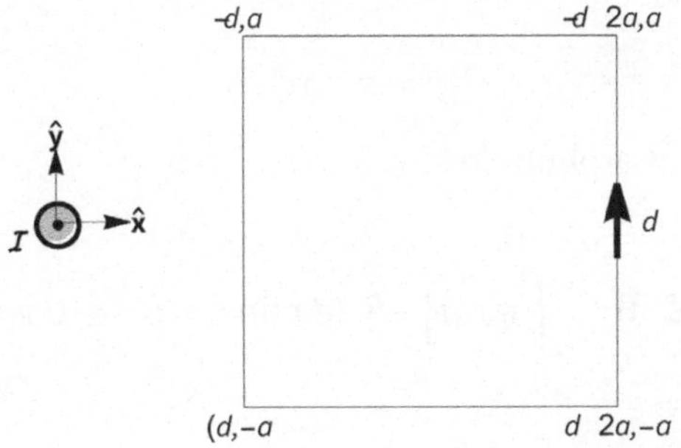

Figure 5.2 The integration path for Ex. 5.3 is a square loop of arbitrary size ($2a$) and arbitrary distance (d) that does not enclose the current.

Example 5.3 Integrate the magnetic field around a square loop of arbitrary side ($2a$) and arbitrary distance (d) that does not enclose the current.

```
In[4]:= $Assumptions = {d > 0, a > 0, x ∈ ℝ, y ∈ ℝ};
        B = {-y I μ0/(2 π (x^2 + y^2)), x I μ0/(2 π (x^2 + y^2)), 0};
        f[y_] = ∫(B[[2]] /. x → d + 2 a) ⅆy;
        g[x_] = ∫(B[[1]] /. y → a) ⅆx ;
        h[y_] = ∫(B[[2]] /. x → d) ⅆy;
        j[x_] = ∫(B[[1]] /. y → -a) ⅆx;
        Simplify[
         (f[a] - f[-a] + g[d] - g[d + 2 a] + h[-a] - h[a] +
            j[d + 2 a] - j[d]) /. d → Random[] a]

Out[9]= 0.
```

5.1.2 Circular Loop

Ampère's law may be illustrated more elegantly with a circular loop as shown in Fig. 5.3. In cylindrical variables, the integration direction is

$$\hat{\boldsymbol{\phi}} = -\sin\phi\,\hat{\mathbf{x}} + \cos\phi\,\hat{\mathbf{y}}$$

(Ex. C.8). It is easiest to integrate around the loop using cartesian coordinates with the angular parameter ϕ. With the origin placed at the wire, the coordinates on the circle are

$$x = d + a\cos\phi$$

and

$$y = a\sin\phi.$$

The line integral of **B** around the circle is

$$\oint d\boldsymbol{\ell}\cdot\mathbf{B} = \int_0^{2\pi} d\phi\, a\left[-B_x(\phi)\sin\phi + B_y(\phi)\cos\phi\right].$$

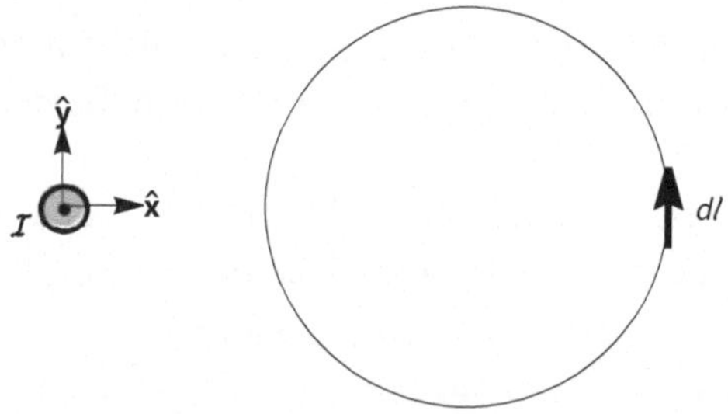

Figure 5.3 The integration path for Ex. 5.4 is a circular loop of radius a with its center placed at an arbitrary distance d along the x-axis. If $d > a$ as shown, the current is not enclosed by the loop. For $d < a$, the current is enclosed.

Example 5.4 Integrate **B** around a circular loop of arbitrary radius a whose center is at a distance d from the origin.

```
In[10]:= $Assumptions = {d > 0, a > 0, x ∈ ℝ, y ∈ ℝ};
         B = {- y I μ0/(2 π (x^2 + y^2)), x I μ0/(2 π (x^2 + y^2)), 0};
         x = d + a Cos[ϕ];
         y = a Sin[ϕ] ;
         a Integrate[-B[[1]] Sin[ϕ] + B[[2]] Cos[ϕ], {ϕ, 0, 2 π}]

Out[13]= 1/2 I (1 + Sign[a - d]) μ0
```

If $a < d$, the charge is not enclosed and the answer is 0. If $a > d$, the charge is enclosed and the answer is $\mu_0 I$.

5.2 DERIVATION FROM THE BIOT-SAVART LAW

Ampére's law is obtained for a steady current by taking the curl of the Biot-Savart law (compare to the relationship between the formula for electric field and Gauss's law as discussed in Sect. 2.3 where the divergence of the electric field was needed). In this case, it means taking the curl of a cross-product,

$$\nabla \times (\mathbf{J} \times \boldsymbol{\mathcal{R}}) = (\boldsymbol{\mathcal{R}} \cdot \nabla)\mathbf{J} - (\mathbf{J} \cdot \nabla)\boldsymbol{\mathcal{R}} + \mathbf{J}\nabla \cdot \boldsymbol{\mathcal{R}} - \boldsymbol{\mathcal{R}}\nabla \cdot \mathbf{J}.$$

Example 5.5 Verify the vector identity for curl of a cross-product.

```
In[14]:= ClearAll["Global`*"];
         J = {f[x, y, z], g[x, y, z], h[x, y, z]};
         ℛ = {p[x, y, z], q[x, y, z], r[x, y, z]};
         Simplify[∇{x,y,z} × (J × ℛ) ==
            (ℛ[[1]] ∂x J + ℛ[[2]] ∂y J + ℛ[[3]] ∂z J) -
             (J[[1]] ∂x ℛ + J[[2]] ∂y ℛ + J[[3]] ∂z ℛ) + J ∇{x,y,z} · ℛ -
            ℛ ∇{x,y,z} · J]

Out[14]= True
```

There is a trick that will make the derivation easier, which is

$$-\nabla\left(\frac{\mathcal{R}_i}{\mathcal{R}^3}\right) = \nabla'\left(\frac{\mathcal{R}_i}{\mathcal{R}^3}\right),$$

where $\mathcal{R}_i$ is some component of $\boldsymbol{\mathcal{R}}$ ($\mathcal{R}_x, \mathcal{R}_y$ or $\mathcal{R}_z$), and ∇' denotes differentiation w.r.t. the primed variables (remembering $\boldsymbol{\mathcal{R}} = \mathbf{r} - \mathbf{r}'$).

Example 5.6 Verify the gradient relationship.

```
In[15]:= 𝓡 = {p[x - x′, y - y′, z - z′], q[x - x′, y - y′, z - z′],
            r[x - x′, y - y′, z - z′]};
         Simplify[-∇{x,y,z} 𝓡[[1]]/(𝓡.𝓡)^(3/2) == ∇{x′,y′,z′} 𝓡[[1]]/(𝓡.𝓡)^(3/2)]
         Simplify[-∇{x,y,z} 𝓡[[2]]/(𝓡.𝓡)^(3/2) == ∇{x′,y′,z′} 𝓡[[2]]/(𝓡.𝓡)^(3/2)]
         Simplify[-∇{x,y,z} 𝓡[[3]]/(𝓡.𝓡)^(3/2) == ∇{x′,y′,z′} 𝓡[[3]]/(𝓡.𝓡)^(3/2)]

Out[16]= True

Out[17]= True

Out[18]= True
```

An additional vector identity is also needed,

$$\nabla\cdot\left(\frac{\mathcal{R}_i}{\mathcal{R}^3}\mathbf{J}\right)=\frac{\mathcal{R}_i}{\mathcal{R}^3}\nabla\cdot\mathbf{J}+\mathbf{J}\cdot\nabla\left(\frac{\mathcal{R}_i}{\mathcal{R}^3}\right).$$

Example 5.7 Verify that $\nabla\cdot(a\mathbf{J})=a\nabla\cdot\mathbf{J}+\mathbf{J}\cdot\nabla(a)$ for any scalar function a.

```
In[19]:= a = s[x, y, z];
         Simplify[∇{x,y,z} · (a J) == a ∇{x,y,z} · J + J.∇{x,y,z} a]

Out[19]= True
```

Putting this together gives

$$\int dv'\mathbf{J}\cdot\nabla'\left(\frac{\mathcal{R}_i}{\mathcal{R}^3}\right)=\int dv'\nabla'\cdot\left(\mathbf{J}\frac{\mathcal{R}_i}{\mathcal{R}^3}\right)-\int dv'\frac{\mathcal{R}_i}{\mathcal{R}^3}\nabla'\cdot\mathbf{J}.$$

The last term is zero for a steady current. Using the divergence theorem, we may take the integration surface out to infinity to get

$$\int dv'\nabla'\cdot\left(\mathbf{J}\frac{\mathcal{R}_i}{\mathcal{R}^3}\right)=\int d\mathbf{a}'\cdot\left(\frac{\mathcal{R}_i}{\mathcal{R}^3}\mathbf{J}\right)=0.$$

There is only one term left in the curl,

$$\nabla\times\mathbf{B}=\int dv'\mathbf{J}\nabla\cdot\left(\frac{\boldsymbol{\mathcal{R}}}{\mathcal{R}^3}\right)=\mu_0\mathbf{J},$$

where the last step follows from $\nabla\cdot(\boldsymbol{\mathcal{R}}/\mathcal{R}^3)=4\pi\delta^3(\boldsymbol{\mathcal{R}})$ (2.3).

5.3 APPLYING AMPÈRE'S LAW

The application of Ampére's law is analogous to Gauss's law in that to use it to calculate the magnetic field one must know its direction from symmetry and be able to perform the line integral with B as a variable. One then solves for B.

5.3.1 Straight Line of Current

If one integrates around a circle with a long wire passing through at its center (Fig. 5.4), then from symmetry the magnetic field is in the $\hat{\boldsymbol{\phi}}$ direction,

$$\oint d\boldsymbol{\ell} \cdot \mathbf{B} = 2\pi r B,$$

and Ampère's law reads

$$2\pi r B = \mu_0 I,$$

giving an equation for the magnitude B,

$$2\pi r B = \mu_0 I,$$

or

$$B = \frac{\mu_0 I}{2\pi r}.$$

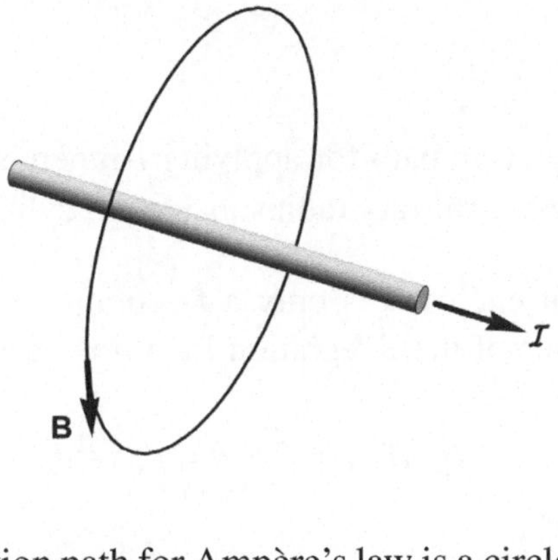

Figure 5.4 The integration path for Ampère's law is a circle of arbitrary radius outside the wire at constant B.

Example 5.8 Calculate the magnetic field 1 cm from the center of a long wire carrying 1 A.

```
In[20]:= r = 1 cm; I = 1 A; N[UnitConvert[μ0 I/(2 π r), T], 2]

Out[20]= 0.000020 T
```

5.3.2 Inside a Long Cylinder of Current

For a cylinder of current (for example, inside a wire) with current density **J** and radius a, one chooses an integration path that is a circle with its center along the cylinder axis and its plane oriented perpendicular to the cylinder axis (Fig. 5.5).

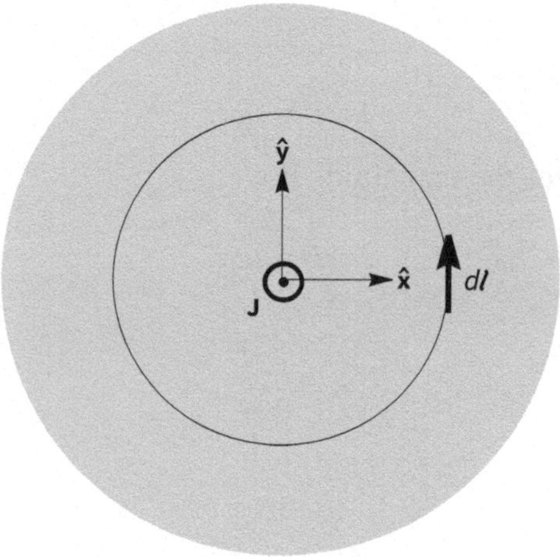

Figure 5.5 The integration path for applying Ampère's law inside a cylinder of current is a circle of arbitrary radius inside the cylinder.

Now the current enclosed is only a fraction of the total current which depends on the radius of the integration loop (the variable r). Ampère's law reads

$$\oint d\boldsymbol{\ell}\cdot\mathbf{B} = 2\pi rB = \mu_0\pi r^2 J,$$

and

$$B = \frac{1}{2}\mu_0 rJ = \frac{\mu_0 rI}{2\pi a^2}.$$

(Note that $I = \pi a^2 J$.)

Example 5.9 Calculate the magnetic field 2 mm from the center of a long wire of radius 4 mm carrying 10 A.

```
In[21]:= a = 4 mm; r = 2 mm; I = 10 A; N[UnitConvert[μ0 r I/(2 π a^2), T], 2]

Out[21]= 0.00025 T
```

5.3.3 Sheet of Current

For a sheet of current density **K**, the integration path is parallel to the sheet on each side with perpendicular connectors (Fig. 5.6).

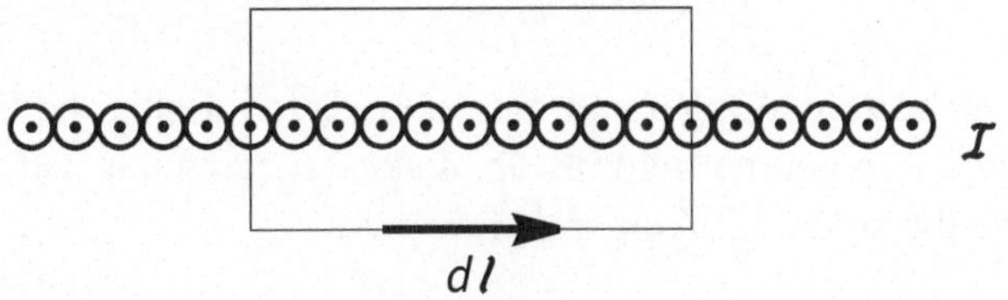

Figure 5.6 The integration path for a current sheet is a rectangular loop where two sides are parallel to the magnetic field.

If each segment parallal to the field has length L, then

$$\oint d\boldsymbol{\ell} \cdot \mathbf{B} = 2LB = \mu_0 KL,$$

and

$$B = \frac{1}{2}\mu_0 K.$$

Example 5.10 Calculate the field for 10 A/m surface current.

```
In[22]:= K = 10 A/m; N[UnitConvert[μ0 K/2, T], 2]

Out[22]= 6.3×10^-6 T
```

5.3.4 Solenoid

In Sect. 4.9 the field of a solenoid on axis was calculated analytically, and the field everywhere for a long solenoid was calculated numerically. Symmetry and Ampère's law can be used to deduce the field of a long solenoid. Take the axis of the solenoid to be the z direction. The current is in the ϕ direction and can only make a field that is in the z direction according to the Biot-Savart law. This can be seen from Ampère's law because an integration loop around ϕ (at constant z) is proportional to B_ϕ but that has zero current enclosed so $B_\phi = 0$, true both inside and outside the solenoid. The field can have no radial component because if we change the direction of the current and turn it upside down, we get the same configuration, again true both inside and out, $B_r = 0$. To get the field outside, make a rectangular integration loop that is in the z direction on two sides and the r direction on the other sides. Since $B_r = 0$, only the path along z can contribute. The line integral of $\mathbf{B}$ (which encloses no current) is

$$\oint d\boldsymbol{\ell} \cdot \mathbf{B} = B_{z_2} L - B_{z_1} L = 0.$$

Since this must be zero along a path that extends to infinity, the field must be zero everywhere. To get the field inside, make a rectangular path in $r - z$ that extends inside the solenoid. Ampère's law gives

$$\oint d\boldsymbol{\ell} \cdot \mathbf{B} = B_z L = \mu_0 N I,$$

where N is the number of turns enclosed by the loop. This result does not depend on the location of the portion of the loop that is inside the solenoid so the field is constant inside. The field depends on the number of turns per length, $n = N/L$,

$$B = \mu_0 n I.$$

Example 5.11 A long solenoid has 10^3 turns per meter. Calculate the current needed to make a 1T field.

```
In[23]:= n = 10^3 m^-1; B = 1 T; N[UnitConvert[B/(μ0 n)], 2]

Out[23]= 8.0×10^2 A
```

5.3.5 Toroid

A toroid has the shape of a solenoid bent into a circle (Fig. 5.7), and like a solenoid, can have any shaped cross-section, although the most common types are circular or rectangular.

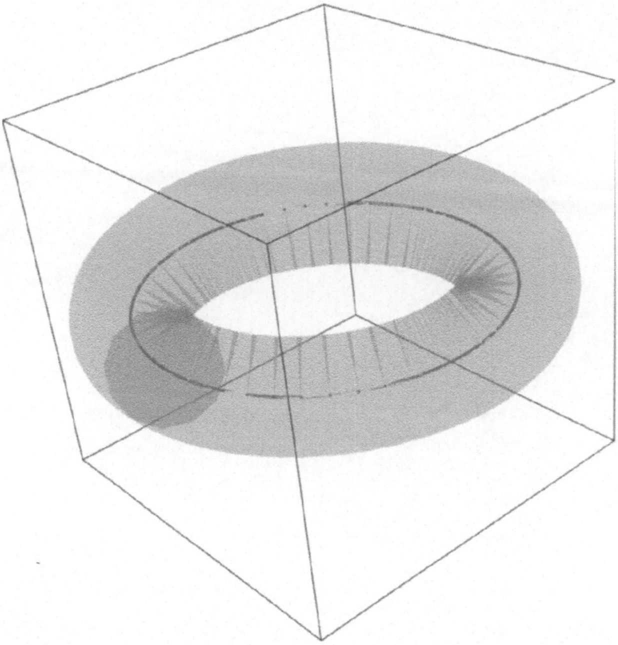

Figure 5.7 A toroid has the shape of a solenoid that is bent into a circle.

The integration path is along the ϕ direction to follow the direction of the magnetic field. The path encloses all the turns of the coil,

$$\oint d\boldsymbol{\ell} \cdot \mathbf{B} = 2\pi r = \mu_0 NI.$$

This gives

$$B = \frac{\mu_0 NI}{2\pi r},$$

where r is the distance to the origin which is placed at the geometrical center (outside the coil). Outside the coil, the field is zero.

5.4 INCOMPLETE MAXWELL EQUATION

The Maxwell equation in differential form is

$$\nabla \times \mathbf{B} = \mu_0 \mathbf{J},$$

and it states that the curl of **B** is zero everywhere except at the location of a current. This rule is true only for steady currents. When the charge density changes with time, there are time-varying electric fields that contribute to the curl of **B**. The complete Maxwell equation is given in Sect. 10.1.

Magnetic Vector Potential

The vector potential **A** is the function whose curl gives the magnetic field,

$$\nabla \times \mathbf{A} = \mathbf{B}.$$

The physical interpretation of **A** is that it represents the momentum per charge in the same way that the electric potential is the potential energy per charge. The units of **A** are $\frac{\text{kg}\cdot\text{m}}{\text{s}\cdot\text{C}} = \text{T}\cdot\text{m}$. Taking the space derivative of **A** with the curl means that the unit of magnetic field must be $\text{kg}\cdot(\text{m/s})$ per C per m, or $\frac{\text{kg}}{\text{s}\cdot\text{C}}$.

Example 6.1 Verify that the units of magnetic field are $\frac{\text{kg}}{\text{s}\cdot\text{C}}$.

```
In[1]:= T == kg/(s C)

Out[1]= True
```

The reason that we can define the magnetic field as the curl of a vector potential is that

$$\nabla \cdot \mathbf{B} = 0,$$

and the divergence of the curl of any vector function is zero. In particular,

$$\nabla \cdot (\nabla \times \mathbf{A}) = 0.$$

Example 6.2 Define an arbitrary vector function and take the divergence of the curl.

```
In[2]:= A = {f[x, y, z], g[x, y, z], h[x, y, z]};
        ∇{x,y,z} · (∇{x,y,z} × A)

Out[3]= 0
```

6.1 DETERMINING A FROM B

It is not easy to get **A** from **B** in general because it is hard to undo a curl. There are a few cases where symmetry makes the calculation easy.

6.1.1 Vector Potential of a Solenoid

The vector potential due to a long solenoid of radius b may be found by using Stokes' theorem (see Fig. 6.1),

$$\int d\mathbf{a} \cdot \nabla \times \mathbf{A} = \oint d\boldsymbol{\ell} \cdot \mathbf{A}.$$

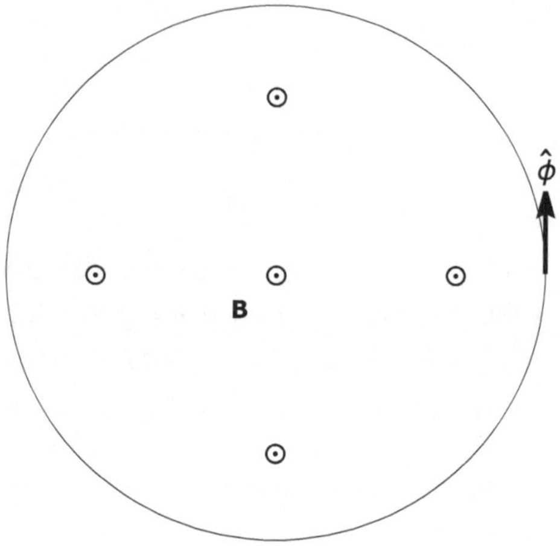

Figure 6.1 A long solenoid has constant field perpendicular to its cross - section. The area integral of the magnetic field over a disk is equal to the line integral of the vector potential around the boundary of the disk.

Knowing the magnetic field to be constant, the surface integral is

$$\int d\mathbf{a} \cdot \nabla \times \mathbf{A} = \int d\mathbf{a} \cdot \mathbf{B} = \pi b^2 B,$$

and the line integral is

$$\oint d\boldsymbol{\ell} \cdot \mathbf{A} = 2\pi b A$$

because the direction of **A** is in the ϕ direction. This gives

$$A = \frac{b}{2} B = \frac{1}{2} \mu_0 n b I$$

for $B = \mu_0 nI$ (4.9 and 5.3.4). Thus,

$$\mathbf{A} = \frac{1}{2}\mu_0 nbI\,\hat{\boldsymbol{\phi}}.$$

One can readily verify that the curl gives **B**.

Example 6.3 Calculate the magnetic field for a long solenoid from the vector potential.

```
In[4]:= Curl[1/2 (μ0 n I r) {0, 1, 0}, {r, ϕ, z}, "Cylindrical"]

Out[4]= {0, 0, n I μ0}
```

6.1.2 Vector Potential of a Sheet of Current

Consider a sheet with surface current density **K** in the x direction. Take the z direction to be perpendicular to the plane of current. The magnetic field is in the y direction (Sect. 5.3.3), so that the vector potential (which is in the x direction) can only depend on z, so

$$\mathbf{A} = f(z)\,\hat{\mathbf{x}},$$

where $f(z)$ is an arbitrary function. Thus,

$$\nabla \times \mathbf{A} = -\frac{\partial f}{\partial z}\,\hat{\mathbf{y}} = \mathbf{B} = \pm\frac{1}{2}\mu_0 K\hat{\mathbf{y}},$$

where the magnetic field is in opposite directions on opposite sides of the current sheet. This gives

$$\mathbf{A} = \pm\frac{1}{2}\mu_0 Kz\,\hat{\mathbf{x}} = \pm\frac{1}{2}\mu_0 z\mathbf{K}.$$

A constant can be added to **A** without changing the magnetic field.

Example 6.4 Verify that this potential is correct for a current sheet by taking the curl.

```
In[5]:= Curl[1/2 μ0 K z {1, 0, 0}, {x, y, z}]

Out[5]= {0, K μ0/2, 0}
```

6.1.3 Vector Potential of a Long Wire

Consider a long wire of radius b. Take the current to be in the z direction. By symmetry, the vector potential can only depend on the distance to the wire r, and one may write

$$\mathbf{A} = f(r)\,\hat{\mathbf{z}}.$$

Outside the wire

$$\nabla \times \mathbf{A} = -\frac{\partial f}{\partial r}\,\hat{\boldsymbol{\phi}} = \frac{\mu_0 I}{2\pi r}\,\hat{\boldsymbol{\phi}}.$$

This gives

$$\mathbf{A} = -\frac{\mu_0 I}{2\pi}\ln\frac{r}{b}\,\hat{\boldsymbol{\phi}}.$$

Example 6.5 Verify that the vector potential outside the wire is correct by taking the curl.

```
In[6]:= Curl[- (μ0 I)/(2 π) Log[r/b] {0, 0, 1}, {r, ϕ, z}, "Cylindrical"]

Out[6]= {0, (I μ0)/(2 π r), 0}
```

Inside the wire

$$\nabla \times \mathbf{A} = -\frac{\partial f}{\partial r}\,\hat{\boldsymbol{\phi}} = \frac{\mu_0 I r}{2\pi b^2}\,\hat{\boldsymbol{\phi}}.$$

This gives

$$\mathbf{A} = -\frac{\mu_0 I}{4\pi b^2}(r^2 - b^2)\,\hat{\boldsymbol{\phi}}.$$

Example 6.6 Verify that the vector potential inside the wire is correct by taking the curl.

```
In[7]:= Curl[- (μ0 I)/(4 π b^2) (r^2 - b^2) {0, 0, 1}, {r, ϕ, z},
          "Cylindrical"]

Out[7]= {0, (r I μ0)/(2 b^2 π), 0}
```

6.2 CALCULATING A BY DIRECT INTEGRATION

The vector potential may be written

$$\mathbf{A} = \frac{\mu_0}{4\pi}\int d\ell'\,\frac{\mathbf{I}}{\mathcal{R}}.$$

It is important to note that **A** is in the same direction as the current. This integral expression for **A** is consistent with the Biot-Savart law. That this is true can be seen by taking the curl with help from the vector identity

$$\nabla \times (f\mathbf{F}) = f\nabla \times \mathbf{F} - \mathbf{F} \times \nabla f.$$

Example 6.7 Show the vector identity holds.

```
In[8]:= a = s[x, y, z];
        Simplify[∇{x,y,z} × (a A) == a ∇{x,y,z} × A - A × ∇{x,y,z} a]

Out[9]= True
```

For the present case,

$$f = \frac{1}{\mathcal{R}},$$

and

$$\mathbf{F} = \mathbf{I}\ell'.$$

The function **F** does not depend on the coordinates (x, y, z) that we are taking the curl with respect to, but rather only on the dummy integration variables, so it has zero curl and one is left with

$$\nabla \times \left(\int d\ell' \frac{\mathbf{I}}{\mathcal{R}} \right) = -\int d\ell' \mathbf{I} \times \nabla \left(\frac{1}{\mathcal{R}} \right) = \int d\ell' \frac{\mathbf{I} \times \boldsymbol{\mathcal{R}}}{\mathcal{R}^3},$$

and the Biot-Savart law is recovered.

Example 6.8 Show that $\nabla\left(\frac{1}{\mathcal{R}}\right) = -\frac{\boldsymbol{\mathcal{R}}}{\mathcal{R}^3}$.

```
In[10]:= ℛ = {x, y, z} - {x′, y′, z′};
                      1                ℛ
         ∇{x,y,z} ─────────── == - ──────────
                    √(ℛ.ℛ)          (ℛ.ℛ)^(3/2)

Out[11]= True
```

For a surface current, the integral for **A** becomes

$$\mathbf{A} = \frac{\mu_0}{4\pi} \int da' \frac{\mathbf{K}}{\mathcal{R}},$$

recalling that **K** is the current per length perpendicular to the direction of current.

For a volume current, the integral for **A** becomes

$$\mathbf{A} = \frac{\mu_0}{4\pi} \int dv' \frac{\mathbf{J}}{\mathcal{R}},$$

recalling that **J** is the current per area perpendicular to the direction of current.

6.2.1 Vector Potential of a Wire Segment

Consider a finite segment of current (Fig. 4.2) for which the magnetic field was calculated using the Biot-Savart law (Ex. 4.9). The vector potential is obtained by direct integration along the path of the current,

$$\mathbf{A} = \frac{\mu_0}{4\pi} \int dz' \frac{I}{\mathcal{R}} \hat{\mathbf{z}}.$$

Example 6.9 Calculate the magnetic vector potential of a line segment extending from z_1 to z_2.

```
In[12]:= ClearAll["Global`*"];
         $Assumptions = { r > 0, z ∈ ℝ};
         ℛ = {r, 0, 0} - {0, 0, z}; f[z_] = (μ0 I)/(4 π) ∫ 1/Sqrt[ℛ.ℛ] ⅆz;
         A = FullSimplify[f[z1] - f[z2]]
```

$$\text{Out[12]=} \frac{I\left(\text{ArcTanh}\left[\frac{z_1}{\sqrt{r^2+z_1^2}}\right] - \text{ArcTanh}\left[\frac{z_2}{\sqrt{r^2+z_2^2}}\right]\right)\mu_0}{4\pi}$$

Calculating **B** from **A** is straightforward. For simple cases, one should be comfortable with visualizing and calculating the curl by hand, while for complicated cases Mathematica is a huge time saver.

In Ex. 6.9 the vector potential a distance r from the axis of an arbitrary line segment was calculated. The direction of **A** is the z direction in cylindrical coordinates. Taking the curl to get **B** is straightforward.

Example 6.10 Calculate the magnetic field from a line segment of current extending from $z_1 = -L/2$ to $z_2 = L/2$ from the vector potential as calculated in ex. 6.9.

```
In[13]:= FullSimplify[
           Curl[A {0, 0, 1}, {r, ϕ, z}, "Cylindrical"] /.
              {z1 → -L/2, z2 → L/2}]
```

$$\text{Out[13]=}\ \left\{0,\ -\frac{L\,I\,\mu_0}{2\,\pi\,r\,\sqrt{L^2+4\,r^2}},\ 0\right\}$$

This is the same answer as that found from the Biot-Savart law in Ex. 4.10.

6.2.2 Vector Potential for a Spinning Sphere of Charge

Consider a sphere of charge of radius R and uniform surface charge density σ that is spinning with uniform angular velocity $\boldsymbol{\omega}$ (Fig. 6.2). The surface current is

$$\mathbf{K} = \sigma\mathbf{v} = \sigma R \sin\theta\omega\,\hat{\boldsymbol{\phi}}.$$

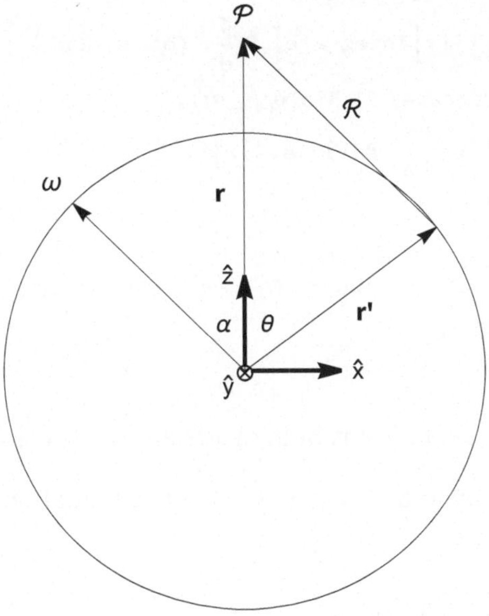

Figure 6.2 A spinning sphere of charge has its rotation axis at an arbitrary angle α, and the angle θ is the polar integration angle.

The direction of the current is in the ϕ direction in a coordinate system where the polar angle is α. This integration is hard in these coordinates. It is much easier to choose the observation point $\mathcal{P}$ along the z-axis and let the angular velocity vector make an arbitrary angle α with the z-axis. Then the integration variable is the ordinary polar angle. After the integration, one can then rotate back to a more natural coordinate system where $\boldsymbol{\omega}$ points along the z-axis. This one is harder to integrate without Mathematica.

The integration is easiest to set up in Cartesian coordinates. After the integration, one can convert to spherical coordinates. Since there is ϕ symmetry, one may choose $y = 0$ without loss of generality. The current and vector potential then point in the y direction, which becomes the ϕ direction after rotating back to the frame where $\boldsymbol{\omega}$ points in the z direction.

Example 6.11 Calculate the magnetic vector potential in a coordinate system where the spinning sphere makes an angle α with the z axis.

```
In[14]:= ClearAll["Global`*"];
         $Assumptions = {R ∈ ℝ, R > 0, σ ∈ ℝ, r ∈ ℝ};
         r′ = FromSphericalCoordinates[{R, θ′, ϕ′}];
         ℛ = {0, 0, r} - r′;
         d = FullSimplify[√(ℛ.ℛ)]; K = σ ω {Sin[α], 0, Cos[α]} × r′;
         f = FullSimplify[Integrate[R K/d, {ϕ′, 0, 2π}]];
         g[θ′_] = Integrate[f R Sin[θ′], θ′];
         Ans = Simplify[μ0/(4π) (g[π] - g[0])]

Out[18]= {0, 1/(6 r^2)
          R σ ω ((r^2 + r R + R^2) Abs[r - R] - (r^2 - r R + R^2) Abs[r + R])
          Sin[α] μ0, 0}
```

The result of Ex. 6.11 holds both inside and outside the sphere.

Example 6.12 Evaluate the magnetic vector potential outside the sphere in spherical coordinates.

```
In[19]:= Simplify[
          A = Ans.{0, 1, 0} {0, 0, 1} /.
            {Abs[r - R] → r - R, Abs[r + R] → r + R, α → -θ}]

Out[19]= {0, 0, R^4 σ ω Sin[θ] μ0 / (3 r^2)}
```

The magnetic field is obtained by taking the curl of the vector potential.

Example 6.13 Calculate the magnetic field outside the sphere in spherical coordinates.

```
In[20]:= Simplify[Curl[{0, 0, -(R^4 σ ω Sin[θ] μ0)/(3 r^2)}, {r, θ, ϕ},
           "Spherical"]]
```

$$\text{Out[20]=}\ \left\{-\frac{2R^4\,\sigma\,\omega\,\mathrm{Cos}[\theta]\,\mu_0}{3r^3},\ -\frac{R^4\,\sigma\,\omega\,\mathrm{Sin}[\theta]\,\mu_0}{3r^3},\ 0\right\}$$

This is a remarkable result. The field outside the spinning sphere of charge is a perfect dipole.

Example 6.14 Evaluate the magnetic vector potential inside the sphere in spherical coordinates.

```
In[21]:= Simplify[
           A = Ans.{0, 1, 0} {0, 0, 1} /.
             {Abs[r - R] → R - r, Abs[r + R] → r + R, α → -θ}]
```

$$\text{Out[21]=}\ \left\{0,\ 0,\ \frac{1}{3}\,r\,R\,\sigma\,\omega\,\mathrm{Sin}[\theta]\,\mu_0\right\}$$

Example 6.15 Evaluate the magnetic field inside the sphere in spherical coordinates.

```
In[22]:= B = Simplify[Curl[{0, 0, 1/3 r R σ ω Sin[θ] μ0},
           {r, θ, ϕ}, "Spherical"]]
```

$$\text{Out[22]=}\ \left\{\frac{2}{3}\,R\,\sigma\,\omega\,\mathrm{Cos}[\theta]\,\mu_0,\ -\frac{2}{3}\,R\,\sigma\,\omega\,\mathrm{Sin}[\theta]\,\mu_0,\ 0\right\}$$

This is just a constant field as can be seen by transforming coordinates from spherical to Cartesian because

$$\hat{\boldsymbol{\phi}} = \cos\theta\,\hat{\mathbf{x}} - \sin\theta\,\hat{\mathbf{y}}.$$

Example 6.16 Transform the magnetic field inside the sphere into Cartesian coordinates.

```
In[23]:= TransformedField["Spherical" → "Cartesian", B,
           {r, θ, ϕ} → {x, y, z}] // Simplify
```

Out[23]= $\left\{0,\ 0,\ \frac{2}{3} R \sigma \omega \mu_0\right\}$

This is a another remarkable result. The field inside the spinning sphere of charge is constant.

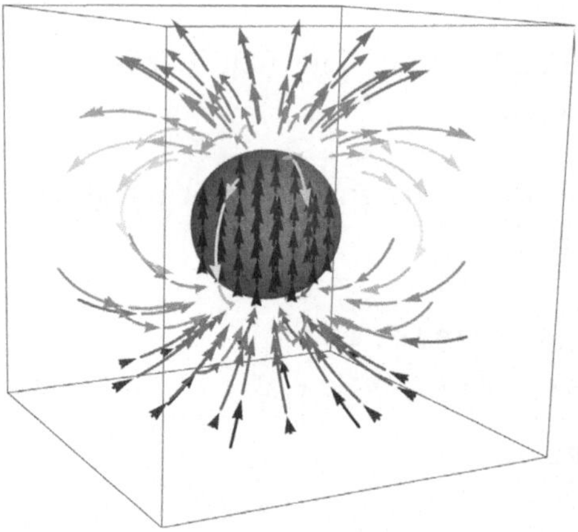

Figure 6.3 The magnetic field from the spinning sphere of charge is constant inside and a perfect dipole outside.

6.2.3 Vector Potential for a Dipole

The ideal dipole field comes from a current loop at large distances. The exact integration of a current loop is not easy. Mathematica will find the limit at large distances with the function AsymptoticIntegrate[$f, x, x \to x_0$] , where the integral $\int dx f$ is performed about $x = x_0$, to any desired accuracy, for the present case in powers of r, the distance to the center of the loop. The integral is set up like normal, and then in a single calculation it will take the limit to any specified order. For a current loop, the $1/r$ term is zero (would correspond to a monopole which does not exist), so the leading term is $1/r^2$. To set up the problem, work in Cartesian coordinates with angular variables. Let the current loop be a circle in the $x-y$ plane centered on the origin with radius b. Since there is ϕ symmetry, take $y = 0$ without loss of generality. The current and **A** are both in the ϕ direction which is the y direction in Cartesian coordinates at $y = 0$. The position coordinate is

$$\mathbf{r} = r\sin\theta\,\hat{\mathbf{x}} + r\cos\theta\,\hat{\mathbf{z}}.$$

The vector that points to the current is

$$\mathbf{r}' = b\cos\phi\,\hat{\mathbf{x}} + b\sin\phi\,\hat{\mathbf{y}}.$$

The current is

$$\mathbf{I} = -I\sin\phi\,\hat{\mathbf{x}} + I\cos\phi\,\hat{\mathbf{y}}.$$

Example 6.17 Calculate the leading term to the vector potential from a circular current loop.

```
In[24]:= ClearAll["Global`*"];
         $Assumptions = {b > 0, ϕ > 0, r > b};
         ℛ = r {Sin[θ], 0, Cos[θ]} - b {Cos[ϕ], Sin[ϕ], 0};
         A = (μ0 I)/(4 π) AsymptoticIntegrate[(b Cos[ϕ])/Sqrt[ℛ.ℛ], {ϕ, 0, 2 π},
             {r, ∞, 2}]

Out[25]= (b^2 I Sin[θ] μ0)/(4 r^2)
```

With the definition of magnetic dipole moment as

$$\mathbf{m} = \pi b^2 I\,\hat{\mathbf{z}},$$

one gets

$$\mathbf{A} = \frac{\mu_0 \mathbf{m}\times\hat{\mathbf{r}}}{4\pi r^2}.$$

The magnetic field is contained by taking the curl.

Example 6.18 Calculate the leading term to the magnetic field from a circular current loop.

```
In[26]:= Curl[(μ0 Sin[θ])/(4 π r^2) m {0, 0, 1}, {r, θ, ϕ}, "Spherical"]

Out[26]= {(m Cos[θ] μ0)/(2 π r^3), (m Sin[θ] μ0)/(4 π r^3), 0}
```

The dipole magnetic field may be written in a coordinate independent form as

$$\mathbf{B} = \frac{\mu_0}{4\pi}\frac{3(\hat{\mathbf{r}}\cdot\mathbf{m})\hat{\mathbf{r}} - \mathbf{m}}{r^3}.$$

Example 6.19 Show that the curl of the magnetic dipole vector potential gives the above expression for **B**.

```
In[27]:= ClearAll["Global`*"]; m = π b² I {0, 0, 1};
         r = {x, y, z};
         Simplify[Curl[μ0 m×r / (4 π (r.r)^(3/2)), {x, y, z}]] ==
           Simplify[μ0/(4 π) (3 (r.m) r / (r.r)^(5/2) - m / (r.r)^(3/2))]

Out[29]= True
```

Example 6.20 A current loop of 1 A and radius 1 cm is oriented in the z direction. Find the vector potential at a position (1 m, 0.5 m, 1 m) from the center of the current loop.

```
In[30]:= m = (1 A) π (1 cm²) {0, 0, 1};
         r = {1, 0.5, 1} m;
         UnitConvert[μ0 m×r / (4 π (r.r)^(3/2)), T m]

Out[30]= {-4.65421×10^-12 mT, 9.30842×10^-12 mT, 0. mT}
```

Example 6.21 Find the magnetic field at a position (1 m, 0.5 m, 1 m) from the center of the current loop.

```
In[31]:= UnitConvert[μ0/(4 π) (3 (r.m) r / (r.r)^(5/2) - m / (r.r)^(3/2)), T]

Out[31]= {1.24112×10^-11 T, 6.20562×10^-12 T, 3.10281×10^-12 T}
```

CHAPTER 7

Faraday's Law

7.1 THE FLUX RULE

The magnetic flux is

$$\Phi = \int da\, \hat{\mathbf{n}} \cdot \mathbf{B} = \int d\mathbf{a} \cdot \mathbf{B}.$$

The flux rule states that a change in magnetic flux through a closed loop introduces an EMF ($\mathcal{E}$) in a direction that opposes the flux change (Lenz's law),

$$\frac{d\Phi}{dt} = -\mathcal{E}.$$

The unit of EMF is the volt and it represents a line integral of the electric field around a closed loop, which unlike the static case, is not zero.

$$\mathcal{E} = \oint d\boldsymbol{\ell} \cdot \mathbf{E},$$

The name EMF comes from "electromotive force," which is a misnomer. It can be useful to think of $\mathcal{E}$ as a potential difference, albeit a special one caused by the changing magnetic flux. The induced $\mathcal{E}$ can be present for either a physical conductor (a wire, for example) or an imaginary loop in empty space. The difference is that if a conducting loop is present, electrons will get pushed by the induced electric field causing a current.

Example 7.1 A magnetic field is perpendicular to a d = 0.1 m square conducting loop and oscillates as (1 T) sin(ωt). Calculate the maximum EMF for for ω = 60 Hz.

```
In[1]:= B = 1 T; d = 0.1 m; ω = 60 / s; UnitConvert[B d² ω, V]

Out[1]= 0.6 V
```

7.2 MAXWELL EQUATION

Faraday's law is a general statement of the flux rule. In integral form, it reads

$$\oint d\boldsymbol{\ell}\cdot\mathbf{E} = -\frac{d}{dt}\int d\mathbf{a}\cdot\mathbf{B}.$$

The differential form comes from Stokes' theorem,

$$\oint d\boldsymbol{\ell}\cdot\mathbf{E} = \int d\mathbf{a}\cdot\nabla\times\mathbf{E},.$$

where the line integral on the left encloses the integration area on the right. This gives the differential form

$$\nabla\times\mathbf{E} = -\frac{\partial\mathbf{B}}{\partial t}.$$

This Maxwell equation is complete and relativistically correct. It pairs with Gauss's law for magnetic fields as the two Maxwell equations that have no source terms (charges or currents):

$$\nabla\cdot\mathbf{B} = 0 \qquad \nabla\times\mathbf{E} + \frac{\partial\mathbf{B}}{\partial t} = 0.$$

For steady currents, we have

$$\nabla\times\mathbf{B} = \mu_0\mathbf{J},$$

which has the solution

$$\mathbf{B} = \frac{\mu_0}{4\pi}\int dv'\frac{\mathbf{J}\times\boldsymbol{\mathcal{R}}}{\mathcal{R}^3}.$$

For the special case of zero charge,

$$\nabla\cdot\mathbf{E} = 0 \qquad \nabla\times\mathbf{E} = -\frac{\partial\mathbf{B}}{\partial t},$$

and the solution is

$$\mathbf{E} = -\frac{1}{4\pi}\frac{\partial}{\partial t}\int dv'\frac{\mathbf{B}\times\boldsymbol{\mathcal{R}}}{\mathcal{R}^3}.$$

Example 7.2 A magnetic field of magnitude 0.005 T oscillates at 60 times per second perpendicular to a circular conducting loop of radius 10 cm. Calculate the EMF around the loop and the maximum electric field inside the conducting loop.

```
In[2]:= ω = 2 π 60 / s; r = 10 cm;
        B = (0.005 T) Sin[ω t];
        Φ = π r^2 B; ℰ = UnitConvert[D[Φ, t]/Cos[ω t], V]

        E = UnitConvert[ℰ/(2 π r), V/m]

Out[2]= 0.0592176 V

Out[3]= 0.0942478 V/m
```

7.3 MUTUAL INDUCTANCE

Two conducting loops with currents I_1 and I_2 are near each other with arbitrary orientation (Fig. 7.1). The magnetic field produced by loop 2 makes a flux in loop 1,

$$\Phi_1 = \int d\mathbf{a}_1 \cdot \mathbf{B}_2 = \int d\mathbf{a}_1 \cdot \nabla \times \mathbf{A}_2 = \oint d\boldsymbol{\ell}_1 \cdot \mathbf{A}_2,$$

where the last step is by Stokes's theorem. Using the direct integral for $\mathbf{A}_2$,

$$\mathbf{A}_2 = \frac{\mu_0 I_2}{4\pi} \int \frac{d\boldsymbol{\ell}_2}{\mathcal{R}},$$

$$\Phi_1 = \frac{\mu_0 I_2}{4\pi} \int \frac{d\boldsymbol{\ell}_1 \cdot d\boldsymbol{\ell}_2}{\mathcal{R}}.$$

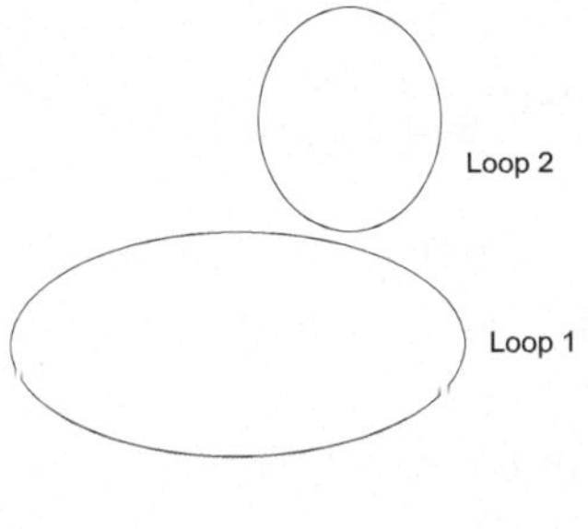

Figure 7.1 Two conducting loops that are near each other have mutual inductance. The magnetic flux through loop 1 (loop 2) is proportional the current in loop 2 (loop 1).

Similarly, the flux in loop 2 caused by loop 1 is

$$\Phi_2 = \frac{\mu_0 I_1}{4\pi} \int \frac{d\boldsymbol{\ell}_1 \cdot d\boldsymbol{\ell}_2}{\mathcal{R}}.$$

Thus, the flux in one of the loops is proportional to the current in the other loop. The constant of proportionality is called the mutual inductance M,

$$M = \frac{\mu_0}{4\pi} \int \frac{d\boldsymbol{\ell}_1 \cdot d\boldsymbol{\ell}_2}{\mathcal{R}}.$$

The unit of inductance is the henry (H). One H is a s·V/A.

Example 7.3 Show 1 H = 1 s·V/A.

```
In[4]:= H == s V/A

Out[4]= True
```

7.3.1 Two-Loop Example

Consider a small loop with radius a and large loop with radius b (fig. 7.2). The loops are parallel and share a common axis, and their centers are separated by a distance d.

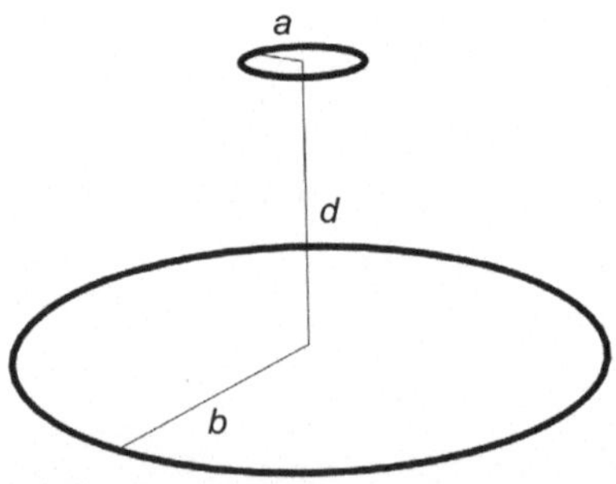

Figure 7.2 The center of a small loop is separated by a distance d from the center of a large loop. The small loop approximates that of a perfect dipole field at the position of the large loop and the large loop approximates a constant field at the position of the small loop.

Example 7.4 Method 1: Take the field from the big loop to be constant at the location of the small loop and calculate the mutual inductance.

```
In[5]:= ClearAll["Global`*"];
        r = {0, 0, d}; r′ = b {Cos[ϕ′], Sin[ϕ′], 0};
        ℛ = r - r′;
```

$$B = \frac{\mu_0}{4\pi}\int_0^{2\pi} b\,\frac{\mathcal{I}\{-\mathrm{Sin}[\phi'],\ \mathrm{Cos}[\phi'],\ 0\}\times\mathcal{R}}{(\mathcal{R}.\mathcal{R})^{3/2}}\,d\phi';$$

$$M = \frac{\pi\, a^2\, B.\{0,\ 0,\ 1\}}{\mathcal{I}}$$

Out[8]=

$$\frac{a^2\, b^2\, \pi\, \mu_0}{2\left(b^2+d^2\right)^{3/2}}$$

Example 7.5 Method 2: Take the field of the small loop to be that of an ideal dipole and calculate the mutual inductance. Note that the flux can be most easily calculated by integrating the radial component of the dipole field over an appropriate portion of a sphere.

In[9]:= `ClearAll["Global`*"];`

$$m = \mathcal{I}\,\pi\, a^2;$$

$$B = \mathrm{Curl}\left[\frac{\mu_0\, \mathrm{Sin}[\theta]}{4\pi r^2}\, m\,\{0,\ 0,\ 1\},\ \{r,\ \theta,\ \phi\},\ \text{"Spherical"}\right];$$

$$\Phi = 2\pi r^2\, \mathrm{Integrate}[B.\{1,\ 0,\ 0\}\ \mathrm{Sin}[\theta],\ \{\theta,\ 0,\ \mathrm{ArcSin}[b/r]\}]\ /.\ r\to \mathrm{Sqrt}\left[b^2+d^2\right];$$

$$M = \frac{\Phi}{\mathcal{I}}$$

Out[11]=

$$\frac{a^2\, b^2\, \pi\, \mu_0}{2\left(b^2+d^2\right)^{3/2}}$$

The results from Ex. 7.4 and Ex. 7.5 agree as expected.

Example 7.6 Calculate the numerical value of the mutual inductance for $a = 0.5$ cm, $b = 10$ cm, and $d = 50$ cm.

```
In[12]:= a = 0.5 cm;
         b = 10 cm; d = 50 cm;
         UnitConvert[M /. μ0 → μ0, H]

Out[12]= 3.72229 × 10^-12 H
```

7.3.2 Nested Solenoids

Consider a smaller solenoid inside a larger solenoid (Fig. 7.3). The flux through the large solenoid is not easy to calculate because of the edge effects. The flux through the small solenoid is straightforward,

$$\Phi = (\mu_0 n_2 I)(\pi a^2)(n_1 d),$$

where n_1 (n_2) is the numbr of turns per length of the smaller (larger) solenoid, a and d are the radius and length of the smaller solenoid, and I is the current in the larger solenoid. The mutual inductance is

$$M = \mu_0 \pi a^2 n_1 n_2 d.$$

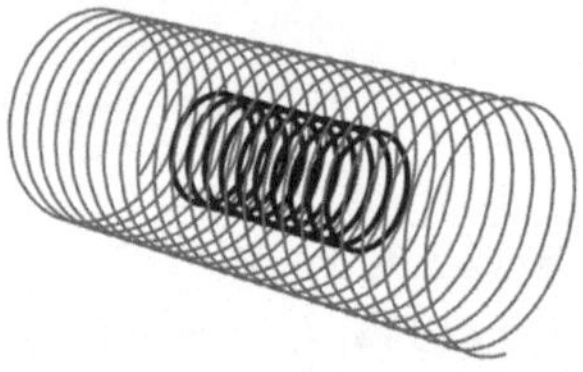

Figure 7.3 Two nested solenoids have a mutual inductance.

Example 7.7 Compute the mutual inductance for nested solenoids with a = 4 cm, d = 16 cm, n_1 = 20 / cm, and n_2 = 10 / cm.

```
In[13]:= a = 4 cm; n1 = 20/cm; n2 = 10/cm; d = 16 cm;
         N[UnitConvert[μ0 π a^2 n1 n2 d, H], 3]

Out[14]= 0.00202 H
```

7.4 SELF-INDUCTANCE

For a single conducting loop, the self-inductance L is defined to be

$$L = \frac{\Phi}{I}.$$

In this case, the current in the loop is making a field that passes though the same loop.

Example 7.8 Calculate the self-inductance of a circular loop with radius 1 m and wire diameter 1 mm. Give a numerical answer in μH.

```
In[15]:= ClearAll["Global`*"]; a = 1; b = 0.001;
         ℛ = {0, r, 0} - {a Cos[ϕ′], a Sin[ϕ′], 0};
         int1[ϕ′_] =
           1/(4 π) Integrate[({-a Sin[ϕ′], a Cos[ϕ′], 0} × ℛ)/(ℛ.ℛ)^(3/2), ϕ′];
         B = int1[2 π] - int1[0]
         UnitConvert[μ0 NIntegrate[2 π r B⟦3⟧, {r, 0, a - b}] m,
           μH]
```

Out[16]=

$$\left\{0, 0, -\left(\left(\left(2 r + (-1 + r)^2 \sqrt{\frac{1 + r^2}{(-1 + r)^2}} \text{EllipticE}\left[\frac{\pi}{4}, -\frac{4 r}{(-1 + r)^2}\right] - (-1 + r^2) \sqrt{\frac{1 + r^2}{(-1 + r)^2}} \text{EllipticF}\left[\frac{\pi}{4}, -\frac{4 r}{(-1 + r)^2}\right]\right) \Big/ \left(4 \pi (-1 + r^2) \sqrt{1 + r^2}\right)\right) + \left(2 r - (-1 + r)^2 \sqrt{\frac{1 + r^2}{(-1 + r)^2}} \text{EllipticE}\left[\frac{3 \pi}{4}, -\frac{4 r}{(-1 + r)^2}\right] + (-1 + r^2) \sqrt{\frac{1 + r^2}{(-1 + r)^2}} \text{EllipticF}\left[\frac{3 \pi}{4}, -\frac{4 r}{(-1 + r)^2}\right]\right) \Big/ \left(4 \pi (-1 + r^2) \sqrt{1 + r^2}\right)\right\}$$

Out[17]= 8.77535 μH

It has been seen that the exact magnetic field due to a current loop is complicated close to the loop. It even depends on the diameter of the wire. The self-inductance of a current loop is a simple conceptual example whose solution is non-trivial.

7.5 STORED MAGNETIC ENERGY

The magnetic energy stored in an inductor (a coil) with current I may be written

$$W = \frac{1}{2}LI^2$$

which corresponds to an EMF

$$\mathcal{E} = -L\frac{dI}{dt}.$$

The flux is

$$\Phi = LI = \int d\mathbf{a}\cdot\mathbf{B} = \int d\mathbf{a}\cdot\nabla\times\mathbf{A} = \oint d\boldsymbol{\ell}\cdot\mathbf{A}.$$

This leads to an alternate expression for the stored energy,

$$W = \frac{1}{2}I\oint d\boldsymbol{\ell}\cdot\mathbf{A} = \frac{1}{2}\oint d\ell\mathbf{A}\cdot\mathbf{I} = \frac{1}{2}\int dv\mathbf{A}\cdot\mathbf{J}.$$

Now use the vector identity

$$\nabla\cdot(\mathbf{A}\times\mathbf{B}) = \mathbf{B}\cdot(\nabla\times\mathbf{A}) - \mathbf{A}\cdot(\nabla\times\mathbf{B}).$$

Example 7.9 Verify the vector identity for divergence of a cross-product.

```
In[18]:= A = {f[x, y, z], g[x, y, z], h[x, y, z]};
         B = {p[x, y, z], q[x, y, z], r[x, y, z]} ;
         Simplify[Div[A×B, {x, y, z}] ==
           B.Curl[A, {x, y, z}] - A.Curl[B, {x, y, z}]]

Out[18]= True
```

Since $\mathbf{B} = \nabla\times\mathbf{A}$ and $\nabla\times\mathbf{B} = \mu_0\mathbf{J}$, one gets the stored energy to be

$$W = \frac{1}{2\mu_0}\left[\int dvB^2 - \int dv\,\nabla\cdot(\mathbf{A}\times\mathbf{B})\right].$$

The second term integrates to zero by the divergence theorem, taking a surface at infinity. The stored energy in a magnetic field is

$$W = \frac{1}{2\mu_0}\int dvB^2.$$

Example 7.10 Find the stored magnetic energy in a length *d* of a coax cable (fig. 7.4) with inner radius *a*, outer radius *b*, and current *I*.

```
In[19]:= ClearAll["Global`*"];
$Assumptions = {a > 0, b > a};
B = μ0 I/(2 π r); W = 1/(2 μ0) 2 π d Integrate[B^2 r, {r, a, b}]
```

Out[19]= $\frac{d I^2 \text{Log}\left[\frac{b}{a}\right] \mu_0}{4 \pi}$

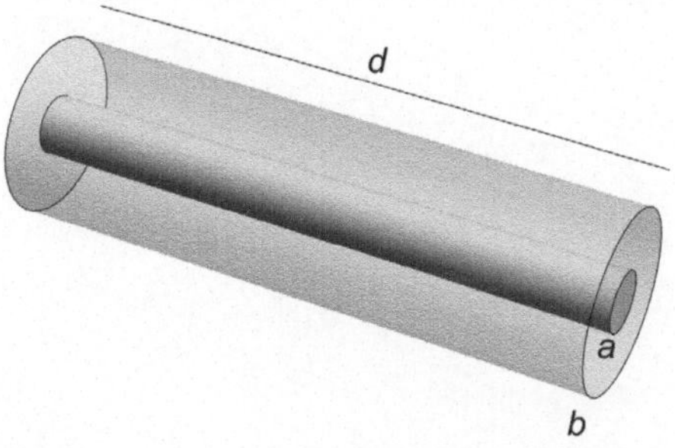

Figure 7.4 A coax cable has its inner conductor at radius *a* and its outer conductor at radius *b*.

Calculating the stored magnetic energy is often the easiest way to calculate the inductance.

Example 7.11 Calculate the inductance in nH for $a = 1$ mm, $b= 4.5$ mm, and $d= 1$ m.

```
In[20]:= L = 2 W/I^2
UnitConvert[
 L /. {a → 1 mm, b → 4.5 mm, d → 1 m, μ0 → μ0}, nH]
```

Out[20]= $\frac{d \text{Log}\left[\frac{b}{a}\right] \mu_0}{2 \pi}$

Out[21]= 300.815 nH

CHAPTER 8

Circuits

8.1 OHM'S LAW

Ohm's law gives the relationship between voltage V and current I in a circuit with resistance R,

$$V = IR.$$

In this expression, V is understood to be the potential difference between two points that have resistance R. The unit of resistance is the ohm which is a volt per amp.

Example 8.1 Verify that 1 Ω= 1 V/A.

```
In[1]:= Ω == V/A

Out[1]= True
```

8.1.1 Electric Field

The resistance can be written in terms of the resistivity ρ and the geometry of the resistor. For a cylinder of length L and cross-sectional area A,

$$R = \frac{\rho L}{A}.$$

Using $J = I/A$, the electric field is

$$E = \frac{V}{L} = \frac{IR}{RA/\rho} = J\rho = \frac{J}{\sigma},$$

where $\sigma = 1/\rho$ is the conductivity. Ohm's law in terms of the field in vector form is

$$\mathbf{J} = \sigma\mathbf{E}.$$

Example 8.2 Get the electrical conductivity (σ) of copper.

```
In[2]:= σ = UnitConvert[ copper ELEMENT [ electrical conductivity ],
            1/(m Ω)]

Out[2]= 5.9×10^7 per meter per ohm
```

It is often stated that the electric field inside a conductor is zero, but this is true only for the static case. For.a steady current (moving charge, there must be an electric field present to cause the force that pushes the charge.

Example 8.3 Calculate the electric field for 1 A in a copper wire with with radius 1 mm.

```
In[3]:= I = 1 A; r = 1 mm; A = π r^2; UnitConvert[(I / A)/σ, V/m]

Out[3]= 0.0054 V/m
```

8.1.2 Drift Speed

The conduction electrons in a metal move randomly at a very high speed that is just below being relativistic. They have a kinetic energy of a few eV. This energy and corresponding speed are referred to as the Fermi energy and Fermi speed (v_F).

Example 8.4 Calculate the Fermi speed for a 4 eV electron in a conductor.

```
In[4]:= K = 4 eV;

        vF = N[UnitConvert[Sqrt[2 K / electron PARTICLE [ mass ]]], 2]

Out[4]= 1.2×10^6 m/s
```

In addition to the large random (Fermi) speed, the electrons acquire a tiny drift speed from the applied electric field. The relationship between the current density **J** and the drift velocity $\mathbf{v}_d$ is

$$\mathbf{J} = \rho \mathbf{v}_\mathrm{d} = ne\mathbf{v}_\mathrm{d},$$

where ρ is the density of electrons that are free to move, typically one per atom, which is expressed as the number density n times the elementary charge e. This gives

$$\mathbf{v}_\mathrm{d} = \frac{J}{ne}.$$

Example 8.5 Calculate the drift speed for 1 A in a wire having a radius of 1 mm.

```
In[5]:= J = A/mm^2; n = 1/(10^-29 m^3); N[UnitConvert[J/(n e)], 2]

Out[5]= 0.000062 m/s
```

The electric current is made up of a huge number of electrons that are moving together very slowly. They are accelerated by the electric field until they suffer a random collision with another electron.

$$\mathbf{v}_\mathrm{d} = a\tau = \frac{eE}{m}\tau = \frac{eE}{m}\frac{d}{\mathbf{v}_\mathrm{d}},$$

where d is the mean free path between collisions. This gives

$$J = ne\mathbf{v}_\mathrm{d} = \frac{eE}{m}\tau = \frac{ne^2 d}{m v_\mathrm{F}} E.$$

This is Ohm's law with

$$\sigma = \frac{ne^2 d}{m v_\mathrm{F}}.$$

Example 8.6 Estimate the mean-free-path for conduction electrons in copper.

```
In[6]:= d = UnitConvert[ copper ELEMENT [ electrical conductivity ]
          electron PARTICLE [ mass ] vF / (n e^2)]

Out[6]= 2.5×10^-8 m
```

Example 8.7 Estimate the time between collisions.

```
In[7]:= UnitConvert[d/vF]

Out[7]= 2.1×10^-14 s
```

8.2 CIRCUITS WITH RESISTORS

8.2.1 Resistors in Series

Resistors in series have the same current. The sum of the potential drops across the individual resistors equals the total potential drop, leading to an equivalent resistance (R_{eq})

$$R_{eq} = R_1 + R_2.$$

8.2.2 Resistors in Parallel

Resistors in parallel have the same potential drop, leading to

$$\frac{1}{R_{eq}} = \frac{1}{R_1} + \frac{1}{R_2},$$

or

$$R_{eq} = \frac{R_1 R_2}{R_1 + R_2}.$$

Example 8.8 Calculate the equivalent resistance for R_1, R_2, R_3 in parallel (Fig. 8.1).

In[8]:= $\texttt{Solve}\left[\frac{1}{R_{eq}} == \frac{1}{R_1} + \frac{1}{R_2} + \frac{1}{R_3}, R_{eq}\right]$

Out[8]= $\left\{\left\{R_{eq} \to \frac{R_1 R_2 R_3}{R_1 R_2 + R_1 R_3 + R_2 R_3}\right\}\right\}$

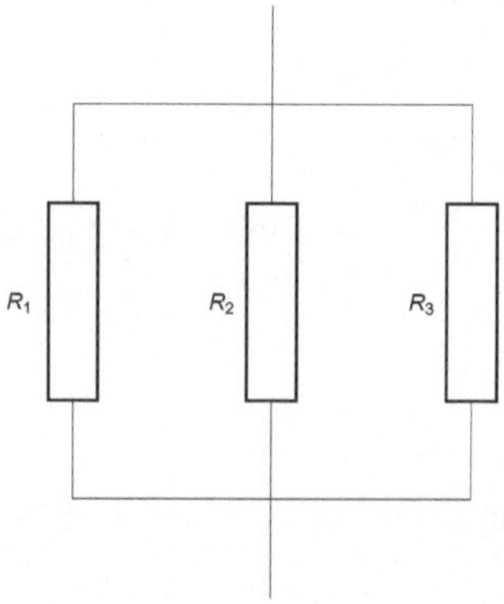

Figure 8.1 Three resistors are shown connected in parallel.

Example 8.9 Calculate the equivalent resistance for R_1, R_2, R_3, R_4 in parallel.

```
In[9]:= Solve[1/Req == 1/R1 + 1/R2 + 1/R3 + 1/R4, Req]

Out[9]= {{Req → (R1 R2 R3 R4)/(R1 R2 R3 + R1 R2 R4 + R1 R3 R4 + R2 R3 R4)}}
```

8.3 KIRCHHOFF'S RULES

The first rule (conservation of charge) says that if we have a junction and the current splits, the net incoming current must equal the net outgoing current.

The second rule states that if we take any closed loop in a circuit, the sum of the voltage drops across each element equals zero. This follows from

$$\oint d\boldsymbol{\ell} \cdot \mathbf{E} = 0.$$

One applies Kirchhoff's rules in a circuit by picking a direction for the current in each branch and then applying junction and loop rules giving simultaneous equations for the currents. The direction picked for the currents does not matter, as the algebraic sign of the solution determines the actual direction.

8.3.1 Two-Loop Circuit

The circuit of Fig. 8.2 has two junctions and three loops (counting the outer loop). One junction and two of the loops give enough information to solve the problem with the other junction and loop providing redundant information. The junction rule at the upper junction gives

$$I_1 = I_2 + I_3.$$

The loop rule on the left gives

$$\mathcal{E}_1 - I_1 R_1 - I_3 R_3 = 0,$$

and on the right gives

$$\mathcal{E}_2 + I_3 R_3 - I_2 R_2 = 0.$$

The bottom junction rule and the outer loop rule do not add any additional information.

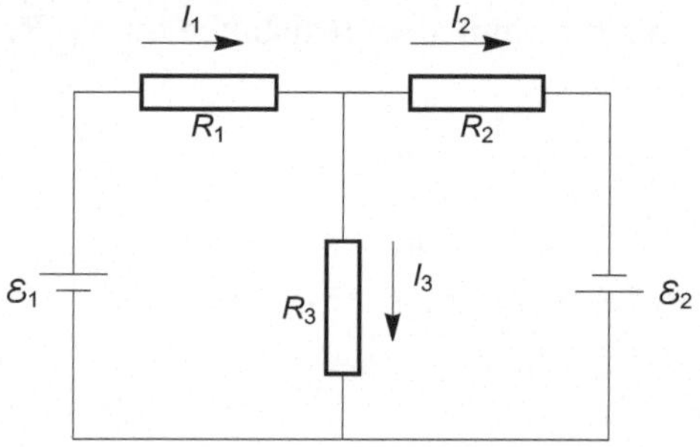

Figure 8.2 The circuit has two loops and a junction which give three independent equations which can be solved for the currents.

Example 8.10 Calculate the currents I_1, I_2, I_3 for the circuit of fig. 8.2.

```
In[10]:= ClearAll["Global`*"];
         Solve[I1 == I2 + I3 && E1 - I1 R1 - I3 R3 == 0 &&
           E2 + I3 R3 - I2 R2 == 0, {I1, I2, I3}] // Simplify
```

Out[10]= $\left\{\left\{I_1 \to \frac{R_2\, \mathcal{E}_1 + R_3\, (\mathcal{E}_1 + \mathcal{E}_2)}{R_2\, R_3 + R_1\, (R_2 + R_3)},\right.\right.$
$\left.\left. I_2 \to \frac{R_1\, \mathcal{E}_2 + R_3\, (\mathcal{E}_1 + \mathcal{E}_2)}{R_2\, R_3 + R_1\, (R_2 + R_3)},\ I_3 \to \frac{R_2\, \mathcal{E}_1 - R_1\, \mathcal{E}_2}{R_2\, R_3 + R_1\, (R_2 + R_3)}\right\}\right\}$

8.3.2 Three-Loop Circuit

The circuit of Fig. 8.3 has two independent junctions and three independent loops out of a total of four junctions and six possible loops. The junction rules

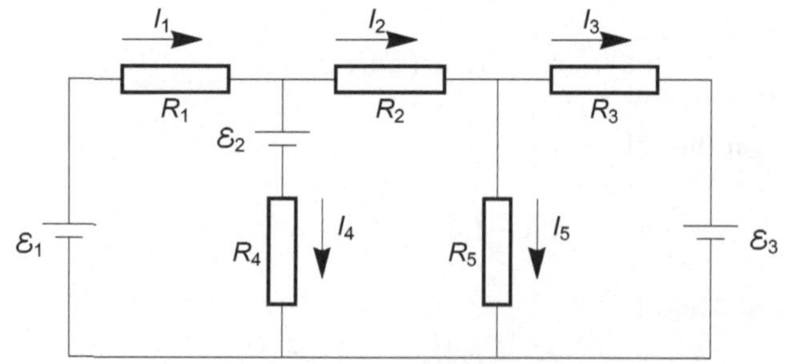

Figure 8.3 The circuit has three loops and two junctions which give five independent equations that can be solved for the currents.

give

$$I_1 = I_2 + I_4$$

and

$$I_2 = I_3 + I_5.$$

The loop rules give

$$\mathcal{E}_1 - I_1 R_1 + \mathcal{E}_2 - I_4 R_4 = 0,$$

$$\mathcal{E}_2 - I_2 R_2 - I_5 R_5 + I_4 R_4 = 0,$$

and

$$-\mathcal{E}_3 + I_5 R_5 - I_3 R_3 = 0,$$

Example 8.11 Calculate the currents I_1, I_2, I_3, I_4, I_5 for the circuit of Fig. 8.3.

```
In[11]:= Solve[I1 == I2 + I4 && I2 == I3 + I5 &&
           ℰ1 - I1 R1 - ℰ2 - I4 R4 == 0 && ℰ2 - I2 R2 - I5 R5 + I4 R4 == 0 &&
           -ℰ3 + I5 R5 - I3 R3 == 0, {I1, I2, I3, I4, I5}] //
          Simplify

Out[11]= {{I1 → (R3 (R4 ℰ1 + R5 (ℰ1 - ℰ2)) + R2 (R3 + R5) (ℰ1 - ℰ2) +
              R4 R5 (ℰ1 - ℰ3)) / (R4 (R3 R5 + R2 (R3 + R5)) +
              R1 (R4 R5 + R2 (R3 + R5) + R3 (R4 + R5))),
          I2 → (R3 (R4 ℰ1 + R1 ℰ2) + R5 (R4 (ℰ1 - ℰ3) + R1 (ℰ2 - ℰ3))) /
            (R4 (R3 R5 + R2 (R3 + R5)) +
              R1 (R4 R5 + R2 (R3 + R5) + R3 (R4 + R5))),
          I3 → (R1 (R5 (ℰ2 - ℰ3) - R2 ℰ3) +
              R4 (R5 (ℰ1 - ℰ3) - (R1 + R2) ℰ3)) /
            (R4 (R3 R5 + R2 (R3 + R5)) +
              R1 (R4 R5 + R2 (R3 + R5) + R3 (R4 + R5))),
          I4 → (R2 (R3 + R5) (ℰ1 - ℰ2) + R3 (R5 (ℰ1 - ℰ2) - R1 ℰ2) +
              R1 R5 (-ℰ2 + ℰ3)) / (R4 (R3 R5 + R2 (R3 + R5)) +
              R1 (R4 R5 + R2 (R3 + R5) + R3 (R4 + R5))),
          I5 → (R3 (R4 ℰ1 + R1 ℰ2) + (R2 R4 + R1 (R2 + R4)) ℰ3) /
            (R4 (R3 R5 + R2 (R3 + R5)) +
              R1 (R4 R5 + R2 (R3 + R5) + R3 (R4 + R5)))}}
```

Example 8.12 Calculate the currents I_1, I_2, I_3, I_4, I_5 for the circuit of Fig. 8.3 for the special case where the resistors are all identical and the batteries are identical.

```
In[12]:= Solve[𝓘₁ == 𝓘₂ + 𝓘₄ && 𝓘₂ == 𝓘₃ + 𝓘₅ &&
           ℰ₁ - 𝓘₁ R₁ - ℰ₂ - 𝓘₄ R₄ == 0 && ℰ₂ - 𝓘₂ R₂ - 𝓘₅ R₅ + 𝓘₄ R₄ == 0 &&
           -ℰ₃ + 𝓘₅ R₅ - 𝓘₃ R₃ == 0, {𝓘₁, 𝓘₂, 𝓘₃, 𝓘₄, 𝓘₅}] /.
         {R₁ → R, R₂ → R, R₃ → R, R₄ → R, R₅ → R, ℰ₁ → ℰ, ℰ₂ → ℰ,
          ℰ₃ → ℰ}
```

Out[12]= $\left\{\left\{\mathcal{I}_1 \to \frac{\mathcal{E}}{8\,R}, \mathcal{I}_2 \to \frac{\mathcal{E}}{4\,R}, \mathcal{I}_3 \to -\frac{3\,\mathcal{E}}{8\,R}, \mathcal{I}_4 \to -\frac{\mathcal{E}}{8\,R}, \mathcal{I}_5 \to \frac{5\,\mathcal{E}}{8\,R}\right\}\right\}$

8.3.3 Δ–Y Transform

This example is a classic, the solution of which is straightforward, but the algebra is tedious if done by hand. Suppose one has a configuration of 3 resistors R_A, R_B, R_C, as shown in Fig. 8.4. What are the values of resistors R_1, R_2, R_3 that make the equivalent circuit of Fig. 8.5? The problem is solved by noting that the points with the potentials V_x and V_y are connected by R_C in parallel with $R_A + R_B$ in FIg. 8.4 and $R_1 + R_2$ in Fig. 8.5. Therefore,

$$\frac{1}{R_1 + R_2} = \frac{1}{R_C} + \frac{1}{R_A + R_B}.$$

Similarly,

$$\frac{1}{R_1 + R_3} = \frac{1}{R_B} + \frac{1}{R_A + R_C}$$

and

$$\frac{1}{R_2 + R_3} = \frac{1}{R_A} + \frac{1}{R_B + R_C}.$$

Example 8.13 Solve for R_1, R_2, R_3

In[13]:= $\text{Solve}\left[\frac{1}{R_1 + R_2} == \frac{1}{R_C} + \frac{1}{R_A + R_B} \;\&\&\; \frac{1}{R_1 + R_3} == \frac{1}{R_B} + \frac{1}{R_A + R_C} \;\&\&\; \frac{1}{R_2 + R_3} == \frac{1}{R_A} + \frac{1}{R_B + R_C}, \{R_1, R_2, R_3\}\right]$

Out[13]= $\left\{\left\{R_1 \to \frac{R_B\,R_C}{R_A + R_B + R_C}, R_2 \to \frac{R_A\,R_C}{R_A + R_B + R_C}, R_3 \to \frac{R_A\,R_B}{R_A + R_B + R_C}\right\}\right\}$

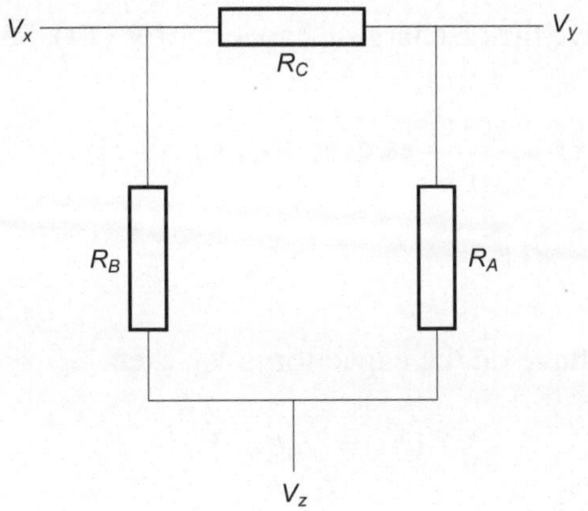

Figure 8.4 Three resistors are shown in the Δ configuration.

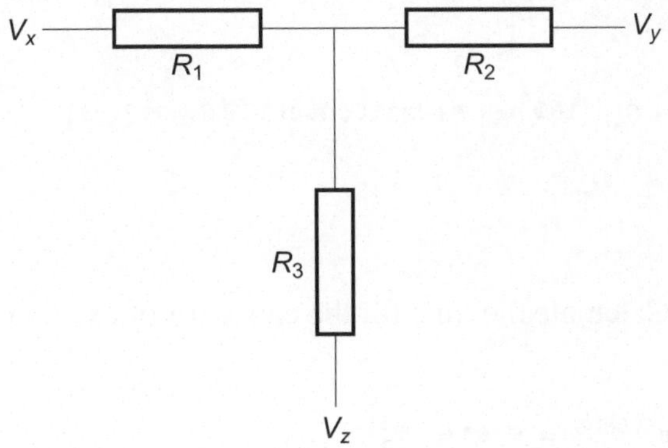

Figure 8.5 Three resistors are shown in the Y configuration.

8.4 RC CIRCUIT

Kirchhoff's loop rule applied to the *RC* circuit gives

$$\frac{Q}{C} - IR = 0.$$

Using $I = -dQ/dt$, one obtains a differential equation for the charge,

$$\frac{dQ}{dt} = -\frac{Q}{RC}.$$

This is a simple differential equation for $Q(t)$ whose solution is an exponential decay.

Example 8.14 Solve the discharging capacitor for $Q(t)$. Set the initial charge equal to q at $t = 0$.

```
In[14]:= DSolve[Q'[t] == -Q[t]/(R C) && Q[0] == q, Q[t], t]

Out[14]= {{Q[t] → e^(-t/(R C)) q}}
```

If the initial voltage on the capacitor is V_0, then

$$Q(t) = V_0 C e^{-\frac{t}{RC}}.$$

The product RC has units of time.

Example 8.15 Calculate the time constant for a 100 pF capacitor discharging with 100 Ω.

```
In[15]:= R = 100 Ω; C = 100 pF; N[UnitConvert[R C, ns], 3]

Out[15]= 10.0 ns
```

Example 8.16 Calculate the time for the capacitor of ex. 8.15 to be 99% discharged.

```
In[16]:= Solve[Q e^(-t/(R C)) == 0.01 Q, t]

Out[16]= {{t → 4.60517 × 10^-8 s}}
```

8.5 ALTERNATING CURENT

Alternating current (AC) circuits have an oscillating, typically sinusoidal, voltage source. This produces an AC current whose phase depends on the circuit elements.

8.6 DRIVEN *LR* CIRCUIT

Figure 8.6 shows an LR circuit with an oscillating voltage source. Kirchhoff's loop rule gives

$$\mathcal{E}_0 \cos \omega t - L\frac{dI}{dt} - RI = 0.$$

The current and voltage are out of phase due to the derivative which turns a cosine (sine) into a minus sine (cosine). The current may be written

$$I = I_0 \cos(\omega t + \phi),$$

where ϕ is the phase angle. The differential equation becomes

$$\mathcal{E}_0 \cos \omega t + \omega L I_0 \sin(\omega t + \phi) - R I_0 \cos(\omega t + \phi) = 0.$$

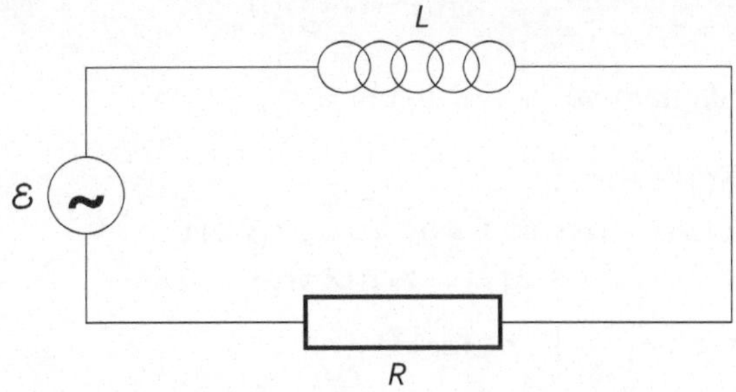

Figure 8.6 The circuit contains an oscillating voltage source, an inductor, and a resistor.

Example 8.17 Solve for the current as a function of time.

```
In[17]:= Clear[R];
         DSolve[ℰ - L I'[t] == R I[t] && I[0] == 0, I[t], t] //
          Simplify
```

$$\text{Out[17]= } \left\{\left\{I[t] \to \frac{\mathcal{E} - e^{-\frac{R t}{L}}\,\mathcal{E}}{R}\right\}\right\}$$

The current has a transient term that decays exponentially. The steady state solution may be obtained by expanding the trigonometric functions.

Example 8.18 Expand the trigonometric functions.

```
In[10]:= TrigExpand[
           ℰ0 Cos[ω t] - L (-I0 ω Sin[ω t + ϕ]) - R I0 Cos[ω t + ϕ] == 0]

Out[18]= Cos[t ω] ℰ0 - R Cos[ϕ] Cos[t ω] I0 + L ω Cos[t ω] Sin[ϕ] I0 +
           L ω Cos[ϕ] Sin[t ω] I0 + R Sin[ϕ] Sin[t ω] I0 == 0
```

Since this must hold for all values of t, the expansion of ex. 8.18 gives 2 equations that may be solved for I_0 and ϕ because the $\sin\omega t$ and $\cos\omega t$ terms must separately be equal to zero. This gives

$$\omega L I_0 \cos\phi + R I_0 \sin\phi = 0,$$

and

$$\mathcal{E}_0 + \omega L I_0 \sin\phi - R I_0 \cos\phi = 0.$$

The solution for ϕ is easy from the first equation,

$$\tan\phi = -\frac{\omega L}{R}.$$

The solution for I_0 involves some algebra.

Example 8.19 Solve for I_0.

```
In[19]:= $Assumptions = {ω > 0, L > 0, R > 0, ℰ > 0};
         Solve[ℰ0 + I0 L ω Sin[ϕ] == I0 R Cos[ϕ] , I0] /.
           ϕ → -ArcTan[L ω / R] // Simplify

Out[19]= {{I0 → ℰ0 / Sqrt[R^2 + L^2 ω^2]}}
```

Example 8.20 Calculate numerical values of I_0 and ϕ for $\mathcal{E}_0 = 110$ V, $\omega =$ 60 Hz, $R = 100\ \Omega$, and $L = 2$ H.

```
In[20]:= ℰ0 = 110 V; ω = 60 Hz; L = 2 H;
         R = 100 Ω; I0 = N[UnitConvert[ℰ0 / Sqrt[R^2 + L^2 ω^2]], 3]
         N[-ArcTan[ω L / R], 3]

Out[20]= 0.704 A

Out[21]= -0.876
```

The current,

$$I = I_0 \frac{\mathcal{E}_0}{\sqrt{R^2 + \omega^2 L^2}} \cos\left(\omega t - \tan^{-1}\frac{\omega L}{R}\right),$$

is seen to peak at a later time than the voltage (fig. 8.7),

$$\mathcal{E} = \mathcal{E}_0 \cos\omega t.$$

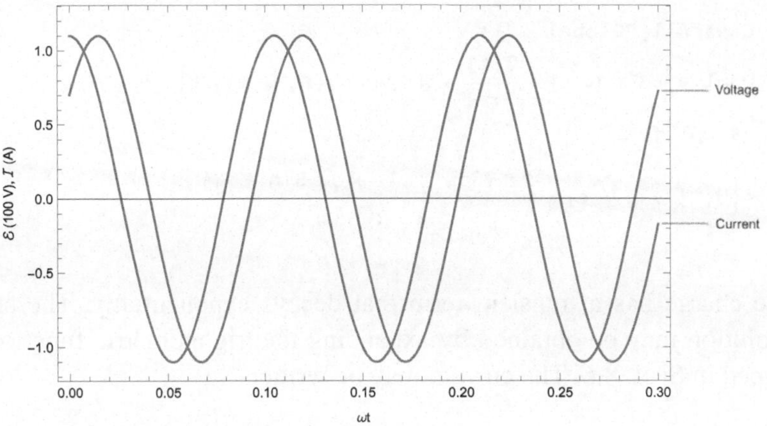

Figure 8.7 The current in an AC *RL* circuit lags the voltage.

8.7 DRIVEN *RC* CIRCUIT

Kirchhoff's loop rule for the RC circuit (Fig. 8.8) gives

$$\mathcal{E}_0 \cos \omega t + \frac{Q}{C} - RI = 0.$$

Using $Q = -dI/dt$,

$$\mathcal{E}_0 \cos \omega t + \frac{Q}{C} + R\frac{dQ}{dt} = 0.$$

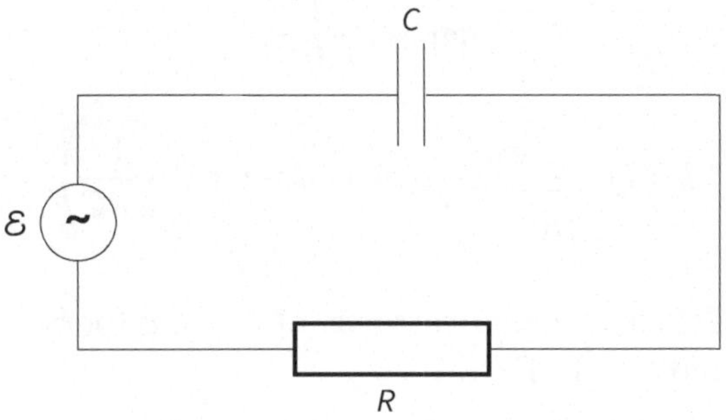

Figure 8.8 The circuit contains an oscillating voltage source, a capacitor, and a resistor.

Example 8.21 Solve for the charge on the capacitor as a function of time.

```
In[22]:= ClearAll["Global`*"];
         DSolve[ℰ Cos[ω t] + Q[t]/C + R Q'[t] == 0, Q[t], t] //
          Simplify

Out[22]= {{Q[t] → e^(-t/(R C)) c1 - (C ℰ (Cos[t ω] + R C ω Sin[t ω]))/(1 + R^2 C^2 ω^2)}}
```

The charge has a transient term that decays exponentially. The steady state solution may be obtained by expanding the trigonometric functions as performed in Sect. 8.6. The current may be written

$$I = I_0 \cos(\omega t + \phi),$$

This gives

$$Q = -\int dt\, I = -\frac{I_0}{\omega}\sin(\omega t + \phi),$$

and

$$\mathcal{E}_0 \cos\omega t - \frac{I_0}{\omega C}\sin(\omega t + \phi) - RI_0\cos(\omega t + \phi) = 0.$$

The solution is identical to the *LR* circuit (8.6) with

$$-\omega L \to \frac{1}{\omega C},$$

giving

$$\tan\phi = \frac{1}{\omega RC},$$

and

$$I = I_0 \frac{\mathcal{E}_0}{\sqrt{R^2 + \frac{1}{\omega^2 C^2}}} \cos\left(\omega t + \tan^{-1}\frac{1}{\omega RC}\right),$$

Example 8.22 Calculate numerical value of I_0 and ϕ for $\mathcal{E}_0 = 110$ V, $\omega = 60$ Hz, $R = 100\ \Omega$, and $C = 1$ nF.

```
In[23]:= ℰ0 = 110 V; ω = 10^6 Hz;  L = 2 H; C = 1 nF;
         R = 100 Ω; I0 = N[UnitConvert[ℰ0/Sqrt[R^2 + (ω L - 1/(ω C))^2], mA], 3]

Out[24]= 0.0550 mA
```

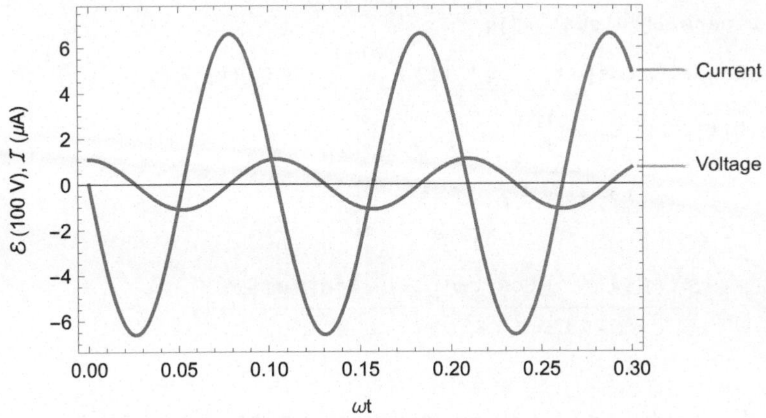

Figure 8.9 The current in an AC *RC* circuit leads the voltage.

8.8 DRIVEN *LRC* CIRCUIT

An *LRC* circuit is shown in Fig. 8.10. Krichhoff's loop rule gives

$$\mathcal{E}_0 \cos \omega t - L\frac{dI}{dt} + \frac{Q}{C} - RI = 0.$$

Using $Q = -dI/dt$,

$$\mathcal{E}_0 \cos \omega t + L\frac{d^2Q}{dt^2} + \frac{Q}{C} + R\frac{dQ}{dt} = 0.$$

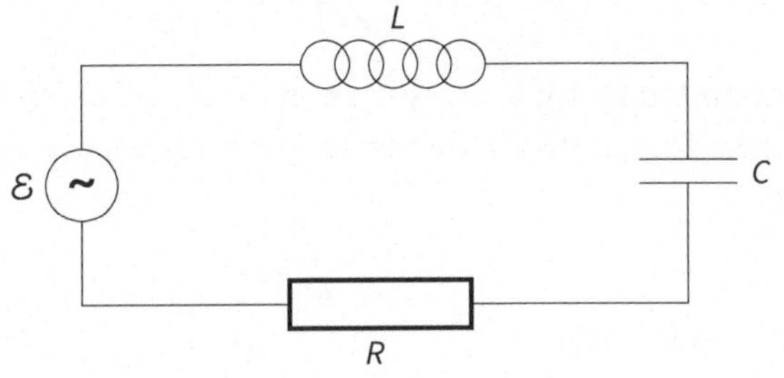

Figure 8.10 The circuit contains an oscillating voltage source, an inductor, a capacitor, and a resistor.

Example 8.23 Solve for the charge on the capacitor as a function of time.

```
In[1]:= ClearAll["Global`*"];
        DSolve[ℰ0 Cos[ω t] - L Q''[t] + Q[t]/C + R Q'[t] == 0,
        Q[t], {t, 0, ∞}]
```

$$\text{Out[1]= } \left\{\left\{Q[t] \to e^{\frac{t\left(R-\frac{\sqrt{4L+R^2 C}}{\sqrt{C}}\right)}{2L}} c_1 + e^{\frac{t\left(R+\frac{\sqrt{4L+R^2 C}}{\sqrt{C}}\right)}{2L}} c_2 - \frac{C\left(\left(1+LC\omega^2\right)\text{Cos}[t\omega]+RC\omega\,\text{Sin}[t\omega]\right)\mathcal{E}_0}{1+2LC\omega^2+R^2C^2\omega^2+L^2C^2\omega^4}\right\}\right\}$$

For the steady-state solution, the current may be written

$$I = I_0 \cos(\omega t + \phi),$$

This gives the capacitor charge and voltage to be

$$Q = -\int dt\, I = -\frac{I_0}{\omega}\sin(\omega t + \phi),$$

$$V_C = \frac{I_0}{\omega C}\sin(\omega t + \phi),$$

and the inductor voltage to be

$$V_L = -i_0 \omega L \sin(\omega t + \phi).$$

The voltage sum is

$$V_C + V_L = -\left(\omega L - \frac{1}{\omega C}\right) I_0 \sin(\omega t + \phi).$$

The capacitor and inductor always have the same phase (both out of phase with the voltage), so for a given frequency, the circuit behaves as an *LR* circuit with inductance

$$\omega L' = \omega L - \frac{1}{\omega C}.$$

From the results of Sect. 8.6,

$$I(t) = \frac{\mathcal{E}_0}{\sqrt{R^2 + \left(\omega L - \frac{1}{\omega C}\right)^2}} \cos(\omega t + \phi),$$

and

$$\tan\phi = \frac{1}{\omega RC}.$$

At large frequencies, the inductor has an important effect on the current.

Example 8.24 Solve for the maximum current and phase.

```
In[2]:= ω = 60 Hz; C = 1 nF;
        R = 100 Ω;
        I0 = N[UnitConvert[ℰ0/Sqrt[R^2 + 1/(ω^2 C^2)]], 3] /. ℰ0 → 110 V
        N[ArcTan[1/(ω R C)], 3]

Out[2]= 6.60×10^-6 A

Out[3]= 1.57
```

CHAPTER 9

Fields Inside Materials

9.1 POLARIZATION VECTOR

The polarization vector $\mathbf{P}$ is the dipole moment per volume from aligned atomic dipoles. The volume integral of $\mathbf{P}$ gives the total dipole vector of the material,

$$\mathbf{p} = \int dv'\, \mathbf{P}.$$

The units of $\mathbf{P}$ are C/m^2, the same as surface charge density.

The polarization might be caused by an external field or it might be permanent. If it is caused by an external field $\mathbf{E}_{\text{ext}}$, the material is referred to as a (linear) dielectric and

$$\mathbf{P} = \varepsilon_0 \chi_e \mathbf{E}_{\text{ext}},$$

where χ_e is the electric susceptibility of the material (covered in Sect. 9.2.1).

Using the dipole potential (Ex. 3.13),

$$V = \frac{\mathbf{p} \cdot \hat{\mathbf{r}}}{4\pi\varepsilon_0 r^2},$$

one can divide the matter into tiny pieces,

$$d\mathbf{p} = \mathbf{P} dv',$$

to get

$$V = \frac{1}{4\pi\varepsilon_0} \int dv' \frac{\mathbf{P} \cdot \hat{\mathcal{R}}}{\mathcal{R}^2}.$$

9.1.1 Bound Charge Density

There is a trick to use to manipulate the integral for the potential into a more useful and intuitive form. Notice that

$$\nabla' \frac{1}{\mathcal{R}} = \frac{\boldsymbol{\mathcal{R}}}{\mathcal{R}^2}.$$

where the differentiation is w.r.t. the primed variables (recall $\mathcal{R} = |\mathbf{r} - \mathbf{r}'|$).

Example 9.1 Verify the vector identity.

```
In[4]:= r = {x, y, z}; r′ = {x′, y′, z′};
        ℛ = r - r′; Grad[1/Sqrt[ℛ.ℛ], {x′, y′, z′}] == ℛ/(ℛ.ℛ)^(3/2)

Out[4]= True
```

This gives

$$V = \frac{1}{4\pi\varepsilon_0}\int dv'\,\mathbf{P}\cdot\left(\nabla'\frac{1}{\mathcal{R}}\right) = \frac{1}{4\pi\varepsilon_0}\int dv'\,\nabla'\cdot\left(\frac{\mathbf{P}}{\mathcal{R}}\right) - \frac{1}{4\pi\varepsilon_0}\int dv'\,\frac{1}{\mathcal{R}}\nabla'\cdot\mathbf{P}.$$

Now use the divergence theorem on the first term on the right to get

$$V = \frac{1}{4\pi\varepsilon_0}\int da'\,\frac{\mathbf{P}\cdot\hat{\mathbf{n}}}{\mathcal{R}} - \frac{1}{4\pi\varepsilon_0}\int dv'\,\frac{\nabla'\cdot\mathbf{P}}{\mathcal{R}}.$$

The first term is a surface charge integral giving the interpretation of the bound surface charge (σ_b) as

$$\sigma_b = \mathbf{P}\cdot\hat{\mathbf{n}},$$

while the second term is a volume charge integral giving the interpretation of the bound volume charge (ρ_b) as

$$\rho_b = -\nabla\cdot\mathbf{P}.$$

Note that the coordinate dependence of **P** must match the differentiation variable.

9.1.2 Radially Polarized Cube

Consider a radially polarized cube (Fig. 9.1) of dimension a,

$$\mathbf{P} = k\mathbf{r}.$$

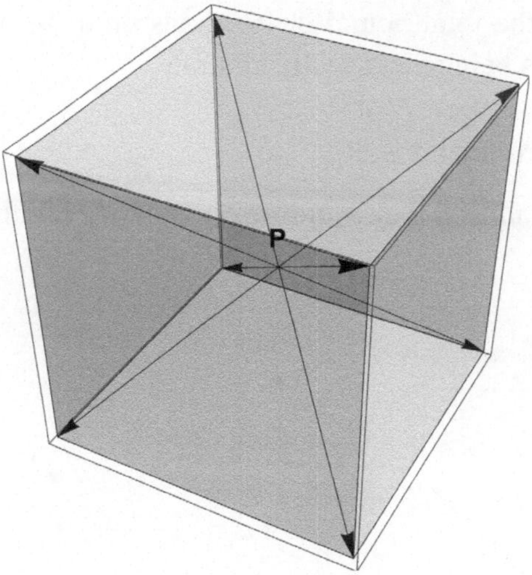

Figure 9.1 A cube is radially polarized.

Example 9.2 Calculate ρ_b for the radially polarized cube.

```
In[11]:= P = k r; -Div[P, {x, y, z}]

Out[11]= -3 k
```

The total bound volume charge is

$$Q_b = -3ka^3.$$

That was easy because the divergence of **r** is just 3. To calculate the bound surface charge is a tad more involved because one must integrate $\mathbf{P} \cdot \hat{\mathbf{n}}$ over the flat surfaces of the cube.

Example 9.3 Calculate the bound surface charge on each face of the cube.

In[12]:= $P = k\left\{x, y, \frac{a}{2}\right\};\ n = \{0, 0, 1\};\ \int_{-a/2}^{a/2}\int_{-a/2}^{a/2} P.n\, dx\, dy$

Out[12]= $\frac{a^3 k}{2}$

The total surface charge is

$$Q_s = 3ka^3$$

and it is seen that the total bound charge adds up to zero as a result of the polarized molecules being electrically neutral.

9.1.3 Uniformly Polarized Ball

Consider a ball of radius a that is uniformly polarized in the z direction (Fig. 9.2).

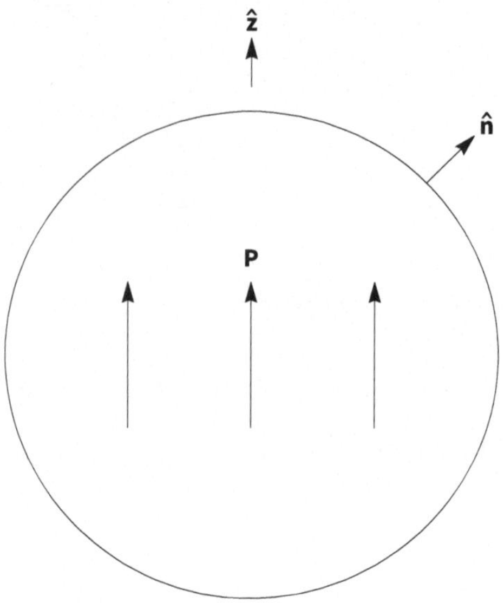

Figure 9.2 A ball is uniformly polarized in the z direction.

The bound volume charge (inside the ball) is zero because $\nabla \cdot \mathbf{P} = 0$. The surface charge may be considered being due to the change in $\mathbf{P}$ from constant to zero at the surface, i.e., $\mathbf{P}$ diverges at the surface. The bound surface charge is

$$\sigma_b = \mathbf{P} \cdot \hat{\mathbf{n}} = \mathbf{P} \cos\theta,$$

where θ is the polar angle (angle w.r.t the z axis). The potential is given by

$$V = \mathbf{P} \cdot \left(\frac{1}{4\pi\varepsilon_0} \int da' \frac{\hat{\boldsymbol{\mathcal{R}}}}{\mathcal{R}^2} \right).$$

The integral in parenthesis is known. It is the same as the integral for the electric field of a ball of charge with constant charge density divided by the charge density, giving $\frac{r}{3\varepsilon_0}\hat{\mathbf{r}}$ inside and $\frac{a^3\mathbf{r}}{3\varepsilon_0 r^3}$ outside, using the calculation of

Sect. 2.4.3 and then dividing by the density. Thus, inside

$$V = \frac{\mathbf{P} \cdot \mathbf{r}}{3\varepsilon_0},$$

and outside

$$V = \frac{a^3 \mathbf{P} \cdot \mathbf{r}}{3\varepsilon_0 r^3}.$$

The electric field is the negative gradient of the potential.

Example 9.4 Calculate the electric field inside the uniformly polarized ball in both spherical and Cartesian coordinates.

```
In[4]:= ClearAll["Global`*"];

         E = -(P/(3 ε0)) Grad[r Cos[θ], {r, θ, ϕ}, "Spherical"]

         TransformedField["Spherical" → "Cartesian", E,
           {r, θ, ϕ} → {x, y, z}] // Simplify

Out[4]= {-(P Cos[θ])/(3 ε0), (P Sin[θ])/(3 ε0), 0}

Out[5]= {0, 0, -P/(3 ε0)}
```

Amazingly, the electric field is uniform inside. It is in the opposite direction as the polarization vector.

Example 9.5 Calculate the electric field outside the uniformly polarized ball.

```
In[6]:= P = (3 p)/(4 π a^3); -((P a^3)/(3 ε0)) Grad[Cos[θ]/r^2, {r, θ, ϕ}, "Spherical"]

Out[6]= {(p Cos[θ])/(2 π r^3 ε0), (p Sin[θ])/(4 π r^3 ε0), 0}
```

This is another amazing result. The electric field outside the ball is that of a perfect electric dipole at the center of the ball, with dipole moment

$$\mathbf{p} = \frac{4}{3}\pi a^3 \mathbf{P}.$$

9.2 DISPLACEMENT VECTOR D

The total charge density ρ may be written as the sum of bound (ρ_b) and free (ρ_f) charge densities,

$$\rho = \rho_\mathrm{b} + \rho_\mathrm{f}.$$

The Maxwell equation for Gauss's law gives

$$\varepsilon_0 \boldsymbol{\nabla} \cdot \mathbf{E} = \rho = \rho_\mathrm{b} + \rho_\mathrm{f} = -\boldsymbol{\nabla} \cdot \mathbf{P} + \rho_\mathrm{f},$$

or

$$\boldsymbol{\nabla} \cdot (\varepsilon_0 \mathbf{E} + \mathbf{P}) = \rho_\mathrm{f}.$$

Defining

$$\mathbf{D} = \varepsilon_0 \mathbf{E} + \mathbf{P},$$

one gets

$$\boldsymbol{\nabla} \cdot \mathbf{D} = \rho_\mathrm{f}.$$

There is a downside using **D** over **E**, even though the Maxwell equation looks simpler, because there is no equivalent of electric potential for **D**. The curl of **D** is not guaranteed to be zero for the static case. In fact,

$$\boldsymbol{\nabla} \times \mathbf{D} = \boldsymbol{\nabla} \times \mathbf{P}.$$

One needs to be very careful using **D**.

9.2.1 Linear Dielectric

A linear dielectric acquires a polarization that is proportional to the applied electric field,

$$\mathbf{P} = \chi_\mathrm{e} \varepsilon_0 \mathbf{E},$$

where the electric susceptibility χ_e is a dimensionless constant.. Then,

$$\mathbf{D} = \varepsilon_0 (1 + \chi_\mathrm{e}) \mathbf{E} = \varepsilon \mathbf{E},$$

where

$$\varepsilon = \varepsilon_0 (1 + \chi_\mathrm{e}).$$

The relative permittivity ε_r is defined as

$$\varepsilon_\mathrm{r} = \frac{\varepsilon}{\varepsilon_0} = (1 + \chi_\mathrm{e}).$$

The effect of the dielectric is to reduce the electric field. For a point charge Q inside a dielectric, Gauss's law reads

$$\mathbf{D} = \frac{Q}{4\pi r^2}\hat{\mathbf{r}},$$

and

$$\mathbf{E} = \frac{\mathbf{D}}{\varepsilon} = \frac{Q}{4\pi\varepsilon r^2}\hat{\mathbf{r}}.$$

9.2.2 Dielectric Ball in an External Electric Field

Consider a uniform dielectric ball in a uniform external field. There is a subtlety in that the field that causes the polarization is the result of the combination of the applied field and the contribution of the field due to the polarization. Let the external field be $\mathbf{E}_0$ and the contribution to field inside the ball caused by the polarization be $\mathbf{E}'_{\text{in}}$. Then the inside field is

$$\mathbf{E}_{\text{in}} = \mathbf{E}_0 + \mathbf{E}'_{\text{in}}.$$

The polarization is

$$\mathbf{P} = \chi_{\text{e}}\varepsilon_0\mathbf{E} = (\varepsilon_{\text{r}} - 1)\varepsilon_0\mathbf{E}_{\text{in}}.$$

Now comes an assumption, which must be verified in the solution, that the resulting field inside is uniform. If the resulting field inside is uniform, then the relationship between $\mathbf{E}'_{\text{in}}$ and $\mathbf{P}$ is known from ex. 9.4,

$$\mathbf{E}'_{\text{in}} = -\frac{\mathbf{P}}{3\varepsilon_0}.$$

This gives

$$\mathbf{E}_{\text{in}} = \mathbf{E}_0 - \frac{\mathbf{P}}{3\varepsilon_0}.$$

Example 9.6 Solve for **P**.

```
In[7]:= ClearAll["Global`*"];
        P = E0 (1/(3 ε0) + 1/((εr - 1) ε0))^-1 // Simplify
```

$$\text{Out[7]=}\ \frac{3\,\varepsilon_0\,(-1+\varepsilon_r)\,E_0}{2+\varepsilon_r}$$

The polarization is indeed uniform and the assumption is verified.

Example 9.7 Solve for $\mathbf{E}_{\text{in}}$.

```
In[8]:=      P
        ─────────────
        (εr - 1) ε0

Out[8]=   3 E0
         ──────
         2 + εr
```

9.3 CAPACITOR WITH A DIELECTRIC

If a capacitor has a dielectric filling, then the electric field is reduced by ε_r and the potential is reduced by the same factor, so the capacitance increases by ε_r . For a parallel plate capacitor (Sect. 3.12.2), $\varepsilon_0 \rightarrow \varepsilon$ and

$$C = \frac{\varepsilon A}{d}.$$

Consider the parallel-plate capacitor of Fig. 9.3. The plate separation is $d_1 + d_2$, and the area is $A_1 = A_2$. The potential difference ΔV is constant. Let the charge density of the left (right) be Q_1/A_1 (Q_2/A_2). The electric fields in the materials ε_1, ε_2, and ε_3 are

$$E_1 = \frac{Q_1}{\varepsilon_1 A_1},$$

$$E_2 = \frac{Q_1}{\varepsilon_2 A_1},$$

and

$$E_3 = \frac{Q_2}{\varepsilon_3 A_2},$$

The potential difference is

$$\Delta V = E_1 d_1 + E_2 d_2 = E_3(d_1 + d_2),$$

or

$$\Delta V = \frac{Q_1 d_1}{\varepsilon_1 A_1} + \frac{Q_1 d_2}{\varepsilon_2 A_1} = \frac{Q_2(d_1 + d_2)}{\varepsilon_3 A_2}.$$

The capacitance is

$$C = \frac{Q_1 + Q_2}{\Delta V} = \frac{Q_1}{\frac{Q_1 d_1}{\varepsilon_1 A_1} + \frac{Q_1 d_2}{\varepsilon_2 A_1}} + \frac{Q_2}{\frac{Q_2(d_1+d_2)}{\varepsilon_3 A_2}}.$$

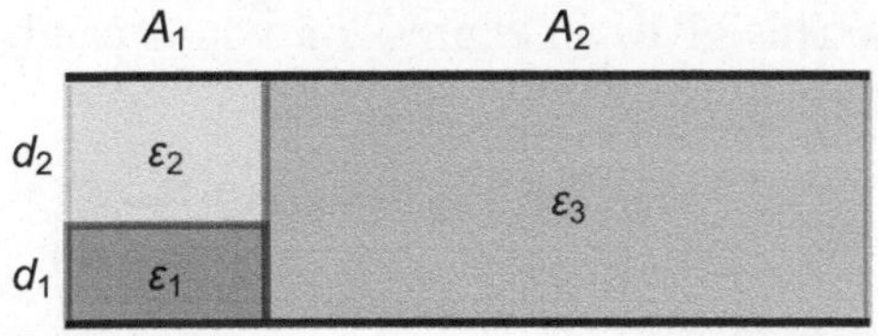

Figure 9.3 A parallel-plate capacitor is filled with three different dielectrics.

or

$$C = \frac{1}{\frac{d_1}{\varepsilon_1 A_1}} + \frac{1}{\frac{d_2}{\varepsilon_2 A_1}} + \frac{1}{\frac{(d_1+d_2)}{\varepsilon_3 A_2}}.$$

This is the result for ε_1 and ε_2 in series and in parallel with ε_3.

Example 9.8 Calculate the capacitance and give a numerical value for $d_1 = 2$ mm, $d_2 = 3$ mm, $A_1 = 1$ cm^2, $A_2 = 4$ cm^2, $\varepsilon_1 = 3\epsilon_0$, $\varepsilon_2 = 2\epsilon_0$, and $\varepsilon_3 = 2.5\epsilon_0$.

```
In[9]:= C1 = ϵ1 A1 / d1; C2 = ϵ3 A1 / d2; C3 = ϵ2 A2 / (d1 + d2);
        Cnet = (1/C1 + 1/C2)^-1 + C3
        UnitConvert[
         Cnet /. {A1 → 1 cm^2, A2 → 4 cm^2, d1 → 2 mm, d2 → 3 mm,
           ϵ1 → 3 ε0, ϵ2 → 2 ε0, ϵ3 → 2.5 ε0}, pF]
```

$$\text{Out[10]=}\quad \frac{A_2 \epsilon_2}{d_1 + d_2} + \frac{1}{\frac{d_1}{A_1 \epsilon_1} + \frac{d_2}{A_1 \epsilon_3}}$$

```
Out[11]= 1.891 pF
```

9.4 POINT CHARGE NEAR A DIELECTRIC BOUNDARY

Consider a point charge Q at a distance d from the a dielectric boundary that forms a plane. The surface bound charge density is

$$\sigma_{\rm b} = \mathbf{P} \cdot \hat{\mathbf{n}} = \varepsilon_0 \chi_{\rm e} E_{z,\rm below},$$

where $E_{z,\rm below}$ is the electric field inside the dielecteic just below the surface. The quantity $E_{z,\rm below}$ is the sum of two contributions, that of the bound surface charge plus that of the point charge. These two contributions point in opposite directions. This gives an expression which can be used to solve for the bound charge density.

$$\sigma_b = \varepsilon_0 \chi_e \left(-\frac{\sigma_b}{2\varepsilon_0} - \frac{Qd}{4\pi\varepsilon_0 (r^2 + d^2)^{3/2}} \right).$$

Example 9.9 Solve for σ_b.

```
In[12]:= ClearAll["Global`*"];
         Solve[σb == ε0 χe (-σb/(2 ε0) - Q d/(4 π ε0 (r^2 + d^2)^(3/2))), σb] //
           Simplify // Factor
```

$$\text{Out[12]=} \left\{\left\{\sigma_b \to -\frac{d\,Q\,\chi_e}{2\pi\left(d^2+r^2\right)^{3/2}(2+\chi_e)}\right\}\right\}$$

Example 9.10 Find the total bound charge.

```
In[13]:= ClearAll["Global`*"];
         $Assumptions = {d > 0, x′ ∈ ℝ, z′ ∈ ℝ};
         Integrate[Integrate[-1/(2 π) χe/(χe + 2) q d/(x′^2 + d^2 + z′^2)^(3/2),
           {x′, -∞, ∞}], {z′, -∞, ∞}]
```

$$\text{Out[13]=} -\frac{q\,\chi_e}{2+\chi_e}$$

The field is shown in Fig. 9.4.

9.5 MAGNETIZATION VECTOR

Magnetized matter is a collection of atomic magnetic dipoles. The magnetization vector **M** is defined to be the magnetic dipole volume density. The volume integral of **M** gives the total dipole vector of the material,

$$\mathbf{m} = \int dv'\, \mathbf{M}.$$

Using the ideal dipole formula for the vector potential (Sect. 6.2.3),

$$\mathbf{A} = \frac{\mu_0 \mathbf{m} \times \hat{\mathbf{r}}}{4\pi r^2},$$

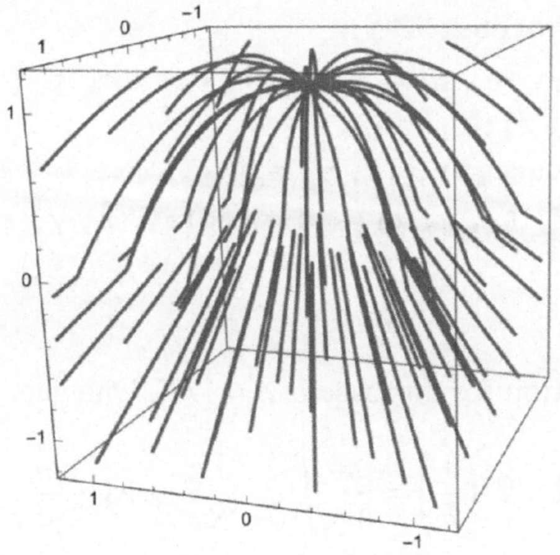

Figure 9.4 The electric field lines from a point charge near a dielectric boundary go from the point charge to the bound charge on the surface outside the dielectric. At the boundary the electric field lines have a kink and are straight lines that point to.the point charge.

the total potential is obtained by superposition,

$$d\mathbf{m} = \mathbf{M}dv',$$

$$d\mathbf{A} = \frac{\mu_0 d\mathbf{m} \times \hat{\mathbf{r}}}{4\pi r^2},$$

and

$$\mathbf{A} = \frac{\mu_0}{4\pi} \int dv' \frac{\mathbf{M} \times \boldsymbol{\mathcal{R}}}{\mathcal{R}^3}.$$

9.5.1 Bound Currents

The calculation is similar to that of manipulating the scalar potential with the polarization vector (Sect. 9.1), but this time one needs the curl of a scalar function times a vector,

$$\boldsymbol{\nabla} \times (a\mathbf{M}) = a\boldsymbol{\nabla} \times \mathbf{M} - \mathbf{m} \times (\boldsymbol{\nabla} a).$$

Example 9.11 Verify the vector identity.

```
In[14]:= ClearAll["Global`*"];
         M = {f[x, y, z], g[x, y, z], h[x, y, z]};
         a = s[x, y, z];
         Simplify[Curl[a M, {x, y, z}] ==
           a Curl[M, {x, y, z}] - M × Grad[a, {x, y, z}]]

Out[15]= True
```

The scalar function for this case is $a = 1/\mathcal{R}$. One gets

$$\mathbf{A} = \frac{\mu_0}{4\pi}\int dv'\mathbf{M}\times\nabla'\left(\frac{1}{\mathcal{R}}\right) = \frac{\mu_0}{4\pi}\int dv'\frac{1}{\mathcal{R}}\nabla'\times\mathbf{M} - \frac{\mu_0}{4\pi}\int dv'\nabla'\times\left(\frac{\mathbf{M}}{\mathcal{R}}\right).$$

The last term can be manipulated with a variation of the divergence theorem. Consider the divergence of a cross product.

$$\nabla\cdot(\mathbf{A}\times\mathbf{B}) = \mathbf{B}\cdot(\nabla\times\mathbf{A}) - \mathbf{A}\cdot(\nabla\times\mathbf{B}).$$

Example 9.12 Verify the vector identity.

```
In[16]:= ClearAll["Global`*"];
         A = {f[x, y, z], g[x, y, z], h[x, y, z]};
         B = {s[x, y, z], t[x, y, z], u[x, y, z]};
         Simplify[Div[A × B, {x, y, z}] ==
           B.Curl[A, {x, y, z}] - A.Curl[B, {x, y, z}]]

Out[17]= True
```

The last identity is applied with $\mathbf{A} = \mathbf{M}/\mathcal{R}$ and $\mathbf{B} = \mathbf{C}$ where $\mathbf{C}$ is a constant vector,

$$\int dv'\nabla'\cdot\left(\frac{\mathbf{M}}{\mathcal{R}}\times\mathbf{C}\right) = \mathbf{C}\cdot\int dv'\nabla'\times\left(\frac{\mathbf{M}}{\mathcal{R}}\right) = \oint d\mathbf{a}'\cdot\left(\frac{\mathbf{M}}{\mathcal{R}}\times\mathbf{C}\right) = \mathbf{C}\cdot\oint d\mathbf{a}'\times\frac{\mathbf{M}}{\mathcal{R}}.$$

Example 9.13 Verify the the last step above is valid, that $\mathbf{a}\cdot(\mathbf{b}\times\mathbf{c}) = \mathbf{c}\cdot(\mathbf{a}\times\mathbf{b})$.

```
In[18]:= a = {e, f, g}; b = {h, j, k}; c = {p, q, r};
         Simplify[a.(b × c) == c.(a × b)]

Out[19]= True
```

Since this result is true for any constant vector **C**, it is true for $\hat{\mathbf{x}}$, $\hat{\mathbf{y}}$, and $\hat{\mathbf{z}}$. Therefore,

$$\int dv' \nabla' \times \left(\frac{\mathbf{M}}{\mathcal{R}} \right) = \oint d\mathbf{a}' \times \frac{\mathbf{M}}{\mathcal{R}}.$$

and

$$\mathbf{A} = \frac{\mu_0}{4\pi} \int dv' \frac{\nabla' \times \mathbf{M}}{\mathcal{R}} + \frac{\mu_0}{4\pi} \oint \frac{\mathbf{M} \times d\mathbf{a}}{\mathcal{R}}.$$

The bound volume current can be identified as

$$\mathbf{J}_{\text{b}} = \nabla \times \mathbf{M},$$

and the bound surface current as

$$\mathbf{K}_{\text{b}} = \mathbf{M} \times \hat{\mathbf{n}}.$$

The vector potential may be written in the convenient form

$$\mathbf{A} = \frac{\mu_0}{4\pi} \int dv' \frac{\mathbf{J}_{\text{b}}}{\mathcal{R}} + \frac{\mu_0}{4\pi} \oint da' \frac{\mathbf{K}_{\text{b}}}{\mathcal{R}}.$$

As in the case of polarization, this is a huge simplification because a complicated integral over atomic dipoles is now reduced to integrals over volume and surface currents.

9.5.2 Uniformly Magnetized Ball

Consider a ball of radius a with uniform magnetization (Fig. 9.5). The bound volume current is zero,

$$\mathbf{J}_{\text{b}} = \nabla \times \mathbf{M} = 0,$$

and the surface current is

$$\mathbf{K}_{\text{b}} = \mathbf{M} \times \hat{\mathbf{n}} = M \sin\theta\, \hat{\boldsymbol{\phi}}.$$

The bound surface current in this problem is the same as form as the free current from a spinning ball of charge (Sect. 6.2.2). The answer is that inside the magnetic field is constant,

$$\mathbf{B} = \frac{2\mu_0}{3} \mathbf{M},$$

and outside it is that of a perfect dipole at the center with dipole vector

$$\mathbf{m} = \frac{4}{4} \pi a^2 \mathbf{M}.$$

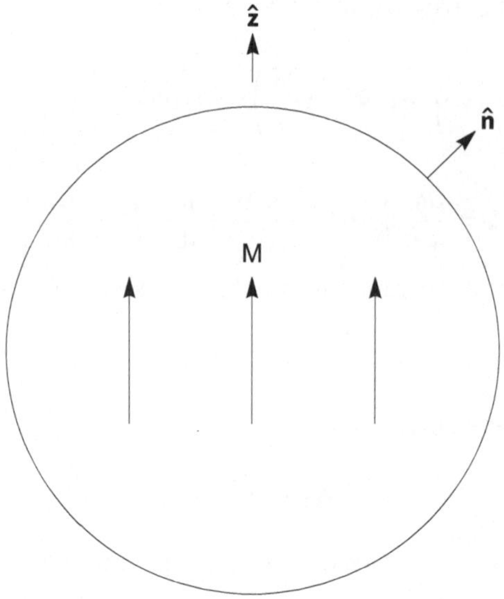

Figure 9.5 A ball is uniformly magnetized in the z direction.

The direct integration for this problem is easy with a little trick. Since **M** is uniform,

$$\mathbf{A} = \frac{\mu_0}{4\pi}\mathbf{M}\times\int dv'\frac{\boldsymbol{\mathcal{R}}}{\mathcal{R}^3}.$$

This integral has been calculated using Gauss's law It is the same one that appears for the electric field of a uniform charge distribution (Sect. 2.4.3). The answer inside was

$$\mathbf{E} = \frac{\rho r}{3\varepsilon_0}\hat{\mathbf{r}} = \frac{1}{4\pi\varepsilon_0}\rho\int dv'\frac{\boldsymbol{\mathcal{R}}}{\mathcal{R}^3}.$$

Therefore,

$$\int dv'\frac{\boldsymbol{\mathcal{R}}}{\mathcal{R}^3} = \frac{4\pi r}{3}\hat{\mathbf{r}},$$

and

$$A = \frac{\mu_0}{4\pi}\mathbf{M}\times\left(\frac{4\pi r}{3}\hat{\mathbf{r}}\right) = \frac{1}{3}\mu_0 M r\sin\theta\,\hat{\boldsymbol{\phi}}.$$

The vector potential makes circles (Fig. 9.6) because the direction of **A** is the same as that of the bound surface current.

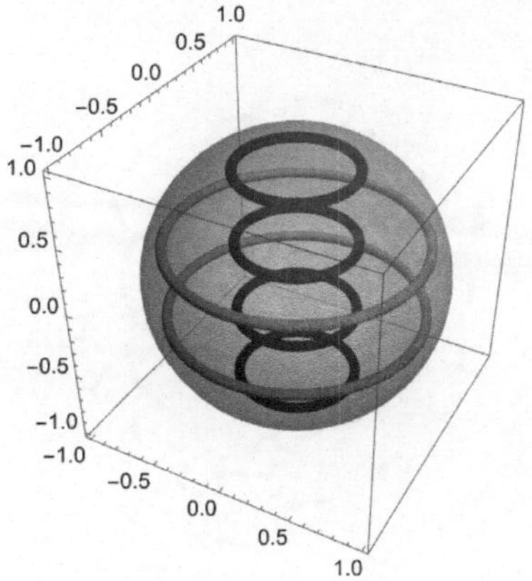

Figure 9.6 The vector potential inside a uniformly magnetized ball makes circles.

Example 9.14 Calculate **B** inside the magnetized ball.

```
In[20]:= ClearAll["Global`*"];
B = Curl[{0, 0, 1/3 μ0 M r Sin[θ]}, {r, θ, ϕ}, "Spherical"]
TransformedField["Spherical" → "Cartesian", B,
  {r, θ, ϕ} → {x, y, z}] // Simplify

Out[20]= {2/3 M Cos[θ] μ0, -2/3 M Sin[θ] μ0, 0}

Out[21]= {0, 0, 2 M μ0/3}
```

Outside the ball,

$$\mathbf{E} = \frac{\frac{4}{3}\pi a^3 \rho}{4\pi\varepsilon_0 r^2}\hat{\mathbf{r}} = \frac{1}{4\pi\varepsilon_0}\rho \int dv' \frac{\boldsymbol{\mathcal{R}}}{\mathcal{R}^3}.$$

(Note that $Q = \frac{4}{3}\pi a^3 \rho$.) Therefore,

$$\int dv' \frac{\boldsymbol{\mathcal{R}}}{\mathcal{R}^3} = \frac{4\pi a^3}{3r^2}\hat{\mathbf{r}},$$

and

$$\mathbf{A} = \frac{\mu_0}{4\pi}\mathbf{M}\times\left(\frac{4\pi a^3}{3r^2}\hat{\mathbf{r}}\right) = \frac{\mu_0 \mathbf{m}\times\hat{\mathbf{r}}}{4\pi r^2}.$$

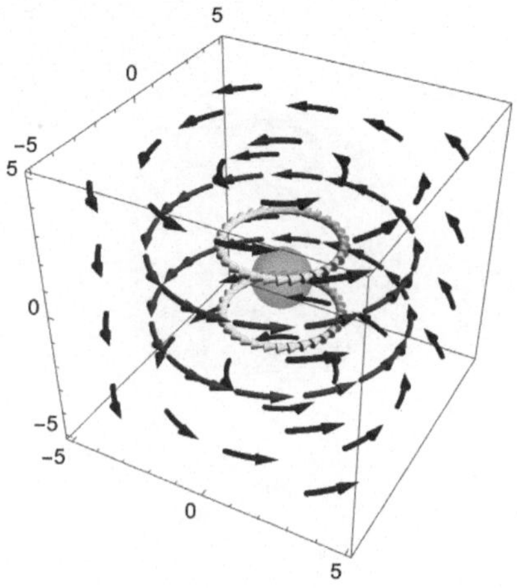

Figure 9.7 The vector potential outside a uniformly magnetized ball makes circles.

(Note that $\mathbf{m} = \frac{4}{3}\pi a^3 \mathbf{M}$.) The vector potential again makes circles (Fig. 9.7).

Example 9.15 Calculate **B** outside the magnetized ball.

In[22]:= $\frac{\mu_0 \, m}{4\pi}$ Curl[{0, 0, $\frac{\text{Sin}[\theta]}{r^2}$}, {r, θ, ϕ}, "Spherical"]

Out[22]= $\left\{\frac{m\,\text{Cos}[\theta]\,\mu_0}{2\pi r^3}, \frac{m\,\text{Sin}[\theta]\,\mu_0}{4\pi r^3}, 0\right\}$

9.6 AUXILIARY FIELD H

The total current density **J** may be written as the sum of bound ($\mathbf{J}_b$) and free ($\mathbf{J}_f$) current densities,

$$\mathbf{J} = \mathbf{J}_b + \mathbf{J}_f.$$

A bound current is illustrated in Fig. 9.8. Ampère's law for steady currents becomes

$$\nabla \times \mathbf{B} = \mu_0(\mathbf{J}_b + \mathbf{J}_f) = \mu_0(\mathbf{J}_b + \nabla \times \mathbf{M}),$$

or

$$\nabla \times \left(\frac{\mathbf{B}}{\mu_0} - \mathbf{M}\right) = \mathbf{J}_f.$$

The auxiliary field **H** is defined by

$$\mathbf{H} = \frac{\mathbf{B}}{\mu_0} - \mathbf{M},$$

which gives

$$\nabla \times \mathbf{H} = \mathbf{J}_{\mathrm{f}}.$$

Unlike **D**, the use of **H** is highly motivated because free currents are measured in the lab. One must be aware, however, that the divergence of **H** is not necessarily zero,

$$\nabla \cdot \mathbf{H} = -\nabla \cdot \mathbf{M}.$$

The auxiliary field **H** diverges at a magnetic boundary, while **B** never diverges.

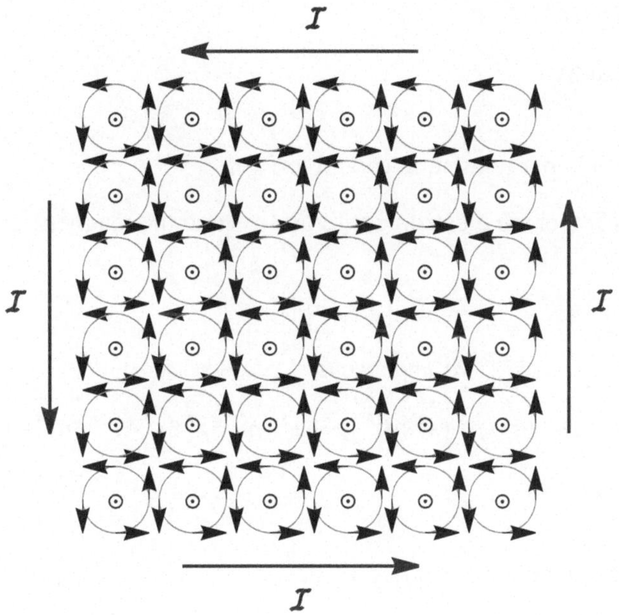

Figure 9.8 A magnetized material has a bound surface current.

9.6.1 Linear Materials

A word of caution is in order in the comparison of the formula for linear materials in the electric and magnetic cases, because historically the magnetic susceptibility χ_{m} is defined in terms of **H** instead of **B**,

$$\mathbf{M} = \chi_{\mathrm{m}}\mathbf{H}.$$

(Compare to $\mathbf{P} = \varepsilon_0 \chi_e \mathbf{E}$.) Another difference is that χ_m is positive for paramagnets and negative for diamagnets, for which there is no electrical analogy. In terms of the field

$$\mathbf{B} = \mu_0(\mathbf{H}+\mathbf{M}) = \mu_0(1+\chi_m)\mathbf{H} = \mu\mathbf{H},$$

where

$$\mu = \mu_0(1+\chi_m).$$

Example 9.16 Get the electron configuration and the magnetic susceptibility for aluminum.

```
In[23]:= aluminum ELEMENT [ electronic configuration ]
         aluminum ELEMENT [ volume magnetic susceptibility ]
```

Out[23]= $1s^2 2s^2 2p^6 3s^2 3p^1$

Out[24]= 0.0000211

Aluminum is paramagnetic (positive χ_m).

Example 9.17 Get the electron configuration and the magnetic susceptibility for gold.

```
In[25]:= gold ELEMENT [ electronic configuration ]
         gold ELEMENT [ volume magnetic susceptibility ]
```

Out[25]= $1s^2 2s^2 2p^6 3s^2 3p^6 4s^2 3d^{10} 4p^6 5s^2 4d^{10} 5p^6 6s^1 4f^{14} 5d^{10}$

Out[26]= -0.0000344

Gold is diamagnetic (negative χ_m)..

9.6.2 Magnetic Ball in an External Magnetic Field

The magnetic ball in an external magnetic field is similar to the electric case (Sect. 9.2.2). It can be solved by a single iteration with the same technique used in the case of polarization in a uniform electric field. An assumption is made that the ball will produce a linear field that will then contribute to the overall field that causes the magnetization. The assumption must be verified to hold true. Take the direction of $\mathbf{M}$ to be in the z direction. The magnetic field inside the ball is

$$\mathbf{B}_{\text{in}} = \mathbf{B}_0 + C\mathbf{B}_0,$$

where the second term is the contribution from magnetization and C is a constant. The magnetization is

$$\mathbf{M} = \chi_{\mathrm{m}}\mathbf{H} = \frac{\chi_{\mathrm{m}}}{\mu_0(1+\chi_{\mathrm{m}})}\mathbf{B}_{\mathrm{in}}.$$

There is a bound surface current

$$\mathbf{K}_{\mathrm{b}} = \mathbf{M}\times\hat{\mathbf{n}} = M\sin\theta\,\hat{\boldsymbol{\phi}}.$$

This is the same current that was found for the spinning sphere of charge (Sect. 6.2.2). The result was a constant field inside, $\frac{2}{3}\mu R\sigma\omega\hat{\mathbf{z}}$ for a current $K = \sigma R\sin\theta\omega$. For the present case, $R\sigma\omega \to M$, confirming the assumption that the magnetization contributes linearly (proportional to $\mathbf{M}$. This gives

$$C\mathbf{B}_0 = \frac{2}{3}\mu_0 M\,\hat{\mathbf{z}} = \frac{2}{3}\mu_0\frac{\chi_{\mathrm{m}}}{\mu_0(1+\chi_{\mathrm{m}})}(\mathbf{B}_0 + C\mathbf{B}_0),$$

Example 9.18 Solve for C.

```
In[27]:= Clear[χm]; Solve[C == 2/3 χm/(1 + χm) (1 + C), C]

Out[27]= {{C → 2 χm/(3 + χm)}}
```

The field inside is

$$\mathbf{B}_{\mathrm{in}} = \mathbf{B}_0 + C\mathbf{B}_0 = \left(1 + \frac{2\chi_{\mathrm{m}}}{3+\chi_{\mathrm{m}}}\right)\mathbf{B}_0,$$

The field outside is a constant field plus that of a perfect dipole as was determined for the spinning sphere of charge.

9.7 BOUNDARY CONDITIONS

The boundary conditions at the junction of two magnetized materials (Fig. 9.9) may be written as

$$H_{\perp,\mathrm{above}} - H_{\perp,\mathrm{below}} = -(M_{\perp,\mathrm{above}} - M_{\perp,\mathrm{below}}),$$

and

$$\mathbf{H}_{\parallel,\mathrm{above}} - \mathbf{H}_{\parallel,\mathrm{below}} = \mathbf{K}_{\mathrm{f}}\times\hat{\mathbf{n}}.$$

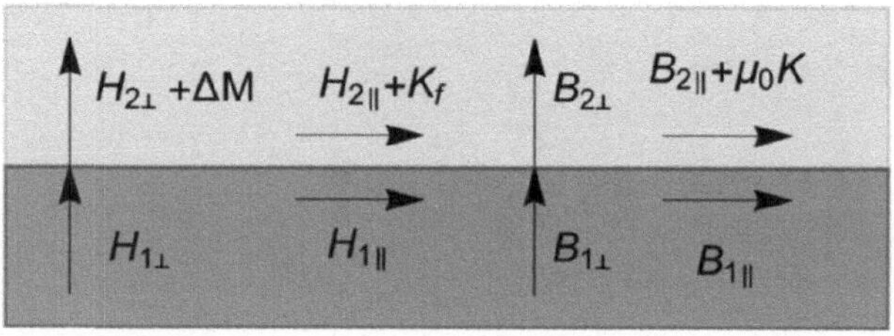

Figure 9.9 The change in **H** at the coundary between 2 materials depends on the **M** and $\mathbf{K}_f$ and the change in **B** depends on the total surface current **K**.

CHAPTER 10

Electromagnetic Waves

10.1 MODIFYING AMPÈRE'S LAW

Taking the divergence of Ampère's law gives

$$\nabla \cdot (\nabla \times \mathbf{B}) = \mu_0 \nabla \cdot \mathbf{J} = 0,$$

since the divergence of the curl of any vector function is zero.

Example 10.1 Calculate the divergence of the curl of an arbitrary function.

```
In[1]:= B = {f[x, y, z], g[x, y, z], h[x, y, z]};
        Div[ Curl[B, {x, y, z}], {x, y, z} ]

Out[1]= 0
```

On the other hand, the continuity equation (4.5) says that

$$\nabla \cdot \mathbf{J} = -\frac{\partial \rho}{\partial t}.$$

This inconsistency can be fixed if a term $\mu_0 \varepsilon_0 \frac{\partial \mathbf{E}}{\partial t}$ is added to the current density term in Ampère's law,

$$\nabla \times \mathbf{B} = \mu_0 \mathbf{J} + \mu_0 \varepsilon_0 \frac{\partial \mathbf{E}}{\partial t}.$$

With this addition the divergence becomes

$$\nabla \cdot (\nabla \times \mathbf{B}) = \mu_0 \nabla \cdot \mathbf{J} + \mu_0 \varepsilon_0 \frac{\partial \nabla \cdot \mathbf{E}}{\partial t} = \mu_0 \nabla \cdot \mathbf{J} + \mu_0 \frac{\partial \rho}{\partial t},$$

where Gauss's law has been used for the divergence of **E**. The right-hand side is now seen to be zero from the continuity equation.

10.2 MAXWELL'S EQUATIONS WITH A MAGNETIC CHARGE

It is a standard exercise to write down what the Maxwell equations would look like if a magnetic charge (monopole) existed. In this case, the equations would need modification by the addition of both a magnetic charge density (ρ_m) and a magnetic current ($\mathbf{J}_m$). The modification should account for conservation of magnetic charge,

$$\nabla \cdot \mathbf{J}_m = -\frac{\partial \rho_m}{\partial t}.$$

The resulting equations are

$$\nabla \cdot \mathbf{E} = \frac{\rho}{\varepsilon_0},$$

$$\nabla \cdot \mathbf{B} = \mu_0 \rho_m,$$

$$\nabla \times \mathbf{E} + \frac{\partial \mathbf{B}}{\partial t} = -\mu_0 \mathbf{J}_m,$$

and

$$\nabla \times \mathbf{B} - \mu_0 \varepsilon_0 \frac{\partial \mathbf{E}}{\partial t} = \mu_0 \mathbf{J}.$$

Conservation of magnetic charge is easily verified by taking the divergence of the modified Faraday's law and substituting the expression for the modified Gauss's law for magnetic fields.

The Lorentz force law becomes

$$\mathbf{F} = q_e(\mathbf{E} + \mathbf{v} \times \mathbf{B}) + q_m(\mathbf{B} - \mathbf{v} \times \mathbf{E}).$$

Since there is a minus sign in the source term for the modified Faraday's law, it follows that the electric $\mathbf{v} \times \mathbf{E}$ term also has this minus sign.

10.3 POYNTING VECTOR

The Poynting vector $\mathbf{S}$ is defined by

$$\mathbf{S} = \frac{1}{\mu_0} \mathbf{E} \times \mathbf{B}.$$

For the case of electromagnetic waves, the Poynting vector "points" in the direction of the wave travel and when time-averaged gives the energy flux, energy per time per area. (Notice that this definition of flux differs from electric or magnetic flux.) Even if there is no electromagnetic radiation, the fields can still have components that are perpendicular to each other, giving a non-zero Poynting vector, meaning that electromagnetic energy is being transported.

The origin of the expression for the Poynting vector comes from energy conservation. Starting with Ampère's law, take the dot product of each term with **E** to get

$$\mathbf{E}\cdot\nabla\times\mathbf{B}-\mu_0\varepsilon_0\mathbf{E}\cdot\frac{\partial\mathbf{E}}{\partial t}=\mu_0\mathbf{E}\cdot\mathbf{J}.$$

This expression may be put in more convenient form by using the identity for divergence of a cross product to replace the first term.

Example 10.2 Verify the vector identity.

```
In[8]:= ClearAll["Global`*"];
        E = {f[x, y, z], g[x, y, z], h[x, y, z]};
        B = {p[x, y, z], q[x, y, z], r[x, y, z]} ;
        Simplify[Div[E × B, {x, y, z}] ==
          B.Curl[E, {x, y, z}] - E.Curl[B, {x, y, z}]]
Out[8]= True
```

Using Faraday's law to eliminate the curl of **E** gives

$$\mathbf{E}\cdot\mathbf{J}=-\frac{1}{2}\frac{\partial}{\partial t}\left(\varepsilon_0E^2+\frac{1}{\mu_0}B^2\right)-\frac{1}{\mu_0}\nabla\cdot(\mathbf{E}\times\mathbf{B}).$$

This expression can be identified with the change in energy as follows. The work done by the fields (dW_i) on a single moving charge q_i in time dt is

$$dW_i=\mathbf{F}\cdot d\boldsymbol{\ell}=q_i(\mathbf{E}+\mathbf{v}\times\mathbf{B})\cdot(\mathbf{v}dt)=q_i\mathbf{E}\cdot\mathbf{v}dt,$$

since $(\mathbf{v}\times\mathbf{B})\cdot\mathbf{v}=0$. Adding up all the charges and substituting $\mathbf{J}=\rho\mathbf{v}$,

$$\frac{dW}{dt}=\sum_i\frac{dW_i}{dt}=\int dv'\,\rho\mathbf{E}\cdot\mathbf{v}+\int dv'\,\mathbf{E}\cdot\mathbf{J}.$$

Thus

$$\frac{dW}{dt}=\mathbf{E}\cdot\mathbf{J}=-\frac{1}{2}\frac{\partial}{\partial t}\left(\varepsilon_0E^2+\frac{1}{\mu_0}B^2\right)-\frac{1}{\mu_0}\nabla\cdot(\mathbf{E}\times\mathbf{B}).$$

The energy per volume u stored in the fields is

$$u=\frac{1}{2}\frac{\partial}{\partial t}\left(\varepsilon_0E^2+\frac{1}{\mu_0}B^2\right).$$

Example 10.3 Calculate the magnetic field that has the same stored energy as an electric field of 1 V/m. Note that $c = 1/\sqrt{\varepsilon_0\mu_0}$.

```
In[3]:= N[UnitConvert[1 V/m / c, nT]]

Out[3]= 3.33564 nT
```

Thus, the statement of local conservation of energy (analogous to the continuity equation) is

$$\nabla \cdot \mathbf{S} = -\frac{\partial u}{\partial t}.$$

The Poynting vector is the "current" for energy density analogous to **J** being the current for charge density.

10.3.1 Current in a Wire

A current-carrying wire (Fig. 10.1) has energy transported in the form of Joule heating. There is an electric field that is in the direction of the current (the field that pushes the charge to make the current) and a magnetic field that curls around the current. The Poynting vector points toward the axis of the wire and describes the flow of electromagnetic energy that appears as Joule heating in the wire. The source of electromagnetic energy is the battery (for example) that causes the current.

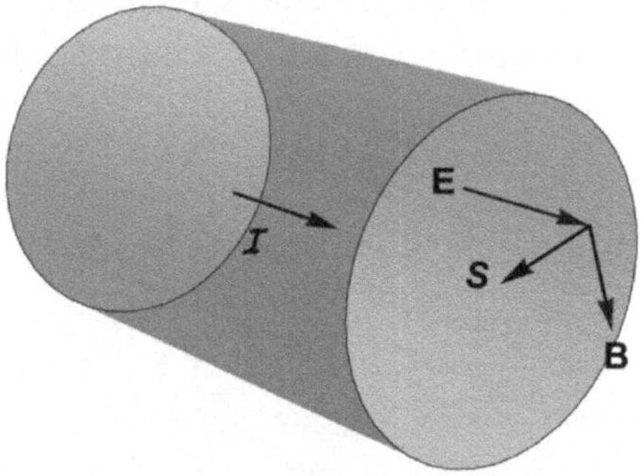

Figure 10.1 A wire carrying a current has an electric field in the direction of the current and a perpendicular magnetic field. The Poynting vector points toward the center of the wire.

For a wire of radius a with current I in the z direction and potential difference V over a length L, the electric field is

$$\mathbf{E} = \frac{V}{L}\,\hat{\mathbf{z}},$$

and the magnetic field is

$$\mathbf{B} = \frac{\mu_0 I}{2\pi a}\hat{\boldsymbol{\phi}}.$$

The Poynting vector is

$$\mathbf{S} = \frac{1}{\mu_0}\mathbf{E}\times\mathbf{B} = \frac{IV}{2\pi aL}\,\hat{\mathbf{r}}.$$

Integrating over the surface of the wire,

$$\oint d\mathbf{a}'\cdot\mathbf{S} = S\,(2\pi aL) = IV.$$

This is the answer expected from Joule heating.

10.3.2 Charging Capacitor

Consider a parallel plate (area A) capacitor that is charging (Fig. 10.2). Taking the z direction to be perpendicular to the plates, the electric field between plates having charge $Q(t)$ is

$$\mathbf{E} = \frac{Q/A}{\varepsilon_0}\,\hat{\mathbf{z}} = \frac{It}{A\varepsilon_0}\,\hat{\mathbf{z}}.$$

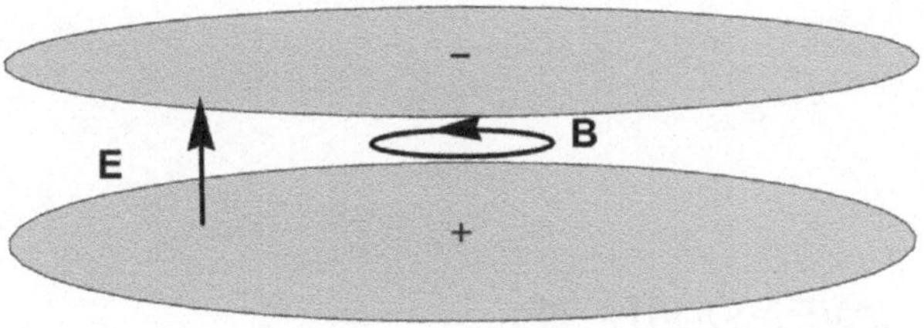

Figure 10.2 A charging capacitor has electric and magnetic fields between the plates that are perpendicular to each other.

The magnetic field between the plates given by Ampère's law,

$$\nabla\times\mathbf{B} = \mu_0\varepsilon_0\frac{\partial\mathbf{E}}{\partial t},$$

or in integral form

$$\oint d\boldsymbol{\ell}'\cdot\mathbf{B} = 2\pi rB = \mu_0\varepsilon_0\frac{\partial(E\pi r^2)}{\partial t} = \mu_0 I.$$

The magnetic field between the plates is

$$\mathbf{B} = \frac{\mu_0 I r}{2A}\,\hat{\boldsymbol{\phi}},$$

the same as the field inside a wire with current I.

Example 10.4 Calculate u and $\mathbf{S}$, and show that $\nabla \cdot \mathbf{S} = -\frac{\partial u}{\partial t}$.

```
In[4]:= ClearAll["Global`*"];
        E = {0, 0, I t/(A ε0)};
        B = {0, μ0 I r/(2 A), 0};
        u = 1/2 ε0 E.E + B.B/(2 μ0)
        S = 1/μ0 E × B
        Div[S, {r, ϕ, z}, "Cylindrical"] == -D[u, t]

Out[5]= t^2 I^2/(2 A^2 ε0) + r^2 I^2 μ0/(8 A^2)

Out[6]= {-r t I^2/(2 A^2 ε0), 0, 0}

Out[7]= True
```

10.4 THE WAVE EQUATION

The Maxwell equations in vacuum are

$$\nabla \cdot \mathbf{E} = 0,$$

$$\nabla \cdot \mathbf{B} = 0,$$

$$\nabla \times \mathbf{E} + \frac{\partial \mathbf{B}}{\partial t} = 0,$$

and

$$\nabla \times \mathbf{B} - \frac{1}{c^2}\frac{\partial \mathbf{E}}{\partial t} = 0,$$

The wave equation for **E** is obtained by taking the curl of Faraday's law,

$$\nabla \times \nabla \times \mathbf{E} + \frac{\partial}{\partial t} \nabla \times \mathbf{B} = \mathbf{0},$$

and then substituting for the curl of **B** using Ampère's law,

$$\nabla \times \nabla \times \mathbf{E} + \frac{1}{c^2}\frac{\partial^2 \mathbf{E}}{\partial t^2} = 0,$$

The double curl has a vector identity

$$\nabla \times \nabla \times \mathbf{E} = \nabla(\nabla \cdot \mathbf{E}) - \nabla^2 \mathbf{E}.$$

Example 10.5 Verify the vector identity.

```
In[2]:= A = {f[x, y, z], g[x, y, z], h[x, y, z]};
        Curl[ Curl[A, {x, y, z}], {x, y, z} ] ==
         Grad[ Div[A, {x, y, z}], {x, y, z} ] -
          Laplacian[A, {x, y, z}]

Out[2]= True
```

Since the divergence of **E** is zero in vacuum,

$$\nabla^2 \mathbf{E} - \frac{1}{c^2}\frac{\partial^2 \mathbf{E}}{\partial t^2} = 0.$$

This is the vector wave equation for **E**; each component of **E** satisfies the wave equation.

The vector wave equation for **B** is identical,

$$\nabla^2 \mathbf{B} - \frac{1}{c^2}\frac{\partial^2 \mathbf{B}}{\partial t^2} = 0.$$

and is obtained by taking the curl of Ampère's law and substituting for the curl of **E** using Faraday's law.

10.5 PLANE WAVES

The simplest wave is the plane wave, so called because the fields are constant at any given time over a plane (usually taken to be the $x-y$ plane). The electric and magnetic fields are in the $x-y$ plane and the wave travels in the direction perpendicular to the plane (taken to be the z direction). The direction of the electric field, referred to as the polarization direction, may be chosen to be any direction in the $x-y$ plane. One can generally write the electric field for the plane wave as

$$\mathbf{E}(z,t) = \mathbf{E}_0 \cos(kz - \omega t).$$

Choosing the x direction, the electric field becomes

$$\mathbf{E}(z,t) = E_0 \cos(kz - \omega t)\,\hat{\mathbf{x}}.$$

The parameter k is called the wave number and it is related to the wavelength λ by

$$k = \frac{2\pi}{\lambda}.$$

Thus, for $z \to z+\lambda$ the electric field is not changed. The parameter ω is the angular frequency which can be written in terms of the period T as

$$\omega = \frac{2\pi}{T} = 2\pi f.$$

where $f = 1/T$. Thus, for $t \to t+T$ the electric field is not changed. Applying the wave equation to the electric field, it is seen to hold true provided that

$$\frac{\omega}{k} = c,$$

which is the same expression as $\lambda f = c$.

Faraday's law gives the direction of **B** relative to **E**.

Example 10.6 Apply Faraday's law to the plane wave.

```
In[9]:= << Notation`
        Symbolize[ParsedBoxWrapper[SubscriptBox["_", "_"]]]
        E = {Ex, Ey, Ez};
        B = {Bx, By, Bz};
        Solve[∇{x,y,z} × (E Cos[k z - ω t]) + ∂t (B Cos[k z - ω t]) == 0,
         B]

Out[11]= {{Bx → -Ey k/ω, By → Ex k/ω, Bz → 0}}
```

Example 10.6 shows that the relationship between **E** and **B** for a plane wave traveling in the z direction is

$$\mathbf{B} = \frac{k}{\omega}\hat{\mathbf{z}} \times \mathbf{E},$$

and that when **E** is in the x direction that **B** must be in the y direction.

10.5.1 Time Average

Time-averaging is useful because electromagnetic waves typically have high frequencies. The time average of the cosine (or sine) over one cycle is 1/2.

Example 10.7 Calculate the time average over 1 period of a time-dependent cosine function with arbitrary phase.

```
In[12]:= 1/T ∫_0^T Cos[k z - 2 π t / T + δ]^2 dt

Out[12]= 1/2
```

10.5.2 Exponential Notation

The fields for electromagnetic plane waves are commonly expressed as complex numbers,

$$\mathbf{E}(z,t) = \mathbf{E}_0 e^{i(kz-\omega t)},$$

and

$$\mathbf{B}(z,t) = \mathbf{B}_0 e^{i(kz-\omega t)},$$

The reason for doing this is that the derivatives which are needed for the Maxwell equations become simple as one does not have to worry about sine going to cosine and vice versa. One just has to remember that the physical fields are the real parts of the complex expressions, noting that

$$e^{i(kz-\omega t)} = \cos(kz-\omega t) + i\sin(kz-\omega t).$$

If the wave direction is an arbitrary direction $\hat{\mathbf{k}}$ and the electric field is in a direction $\hat{\mathbf{n}}$, then

$$\mathbf{E}(\mathbf{r},t) = E_0 e^{i(kz-\omega t)}\, \hat{\mathbf{n}},$$

and

$$\mathbf{B} = \frac{1}{c}\hat{\mathbf{k}} \times \mathbf{E},$$

10.6 WAVES IN MATTER

In the case of no free charge and no free current,

$$\nabla \cdot \mathbf{D} = 0,$$

$$\nabla \cdot \mathbf{B} = 0,$$

$$\nabla \times \mathbf{E} = -\frac{\partial \mathbf{B}}{\partial t},$$

and

$$\nabla \times \mathbf{H} = \frac{\partial \mathbf{D}}{\partial t}.$$

For linear media

$$\mathbf{D} = \varepsilon \mathbf{E},$$

and

$$\mathbf{H} = \frac{1}{\mu}\mathbf{B}.$$

and homogeneous media (ε and μ constant),

$$\nabla \cdot \mathbf{E} = 0,$$

$$\nabla \cdot \mathbf{B} = 0,$$

$$\nabla \times \mathbf{E} = -\frac{\partial \mathbf{B}}{\partial t},$$

and

$$\nabla \times \mathbf{B} = \varepsilon\mu \frac{\partial \mathbf{E}}{\partial t}.$$

The wave speed is

$$v = \frac{1}{\sqrt{\varepsilon\mu}}.$$

The index of refraction n is defined by

$$n = \sqrt{\frac{\varepsilon\mu}{\varepsilon_0\mu_0}}.$$

The expressions for energy density, energy flux, and intensity are recovered with the substitutions $\varepsilon_0 \to \varepsilon$, $\mu_0 \to \mu$, and $c \to v$, giving

$$u = \frac{1}{2}\frac{\partial}{\partial t}\left(\varepsilon E^2 + \frac{1}{\mu}B^2\right),$$

$$\mathbf{S} = \frac{1}{\mu}\mathbf{E} \times \mathbf{B}.$$

The intensity I (energy per area per time) of the wave is

$$I = \frac{1}{2}\varepsilon v E_0^2.$$

The boundary conditions are that parallel E and perpendicular B do not change,

$$E_1^{\|} = E_2^{\|},$$

and

$$B_1^{\perp} = B_2^{\perp}.$$

The boundary conditions on perpendicular E and parallel B are

$$\varepsilon_1 E_1^{\perp} = \varepsilon_2 E_2^{\perp},$$

and

$$\frac{1}{\mu_1} B_1^{\|} = \frac{1}{\mu_2} B_2^{\|}.$$

10.7 REFLECTION AND REFRACTION

Consider a plane wave traveling in medium 1 in the x direction with the electric field aligned along the x direction (Fig. 10.3). Take the wave speed to be v_1. The fields of the incident wave are written

$$\mathbf{E}_{\mathrm{I}} = E_{0\mathrm{I}} e^{i(k_1 z - \omega t)} \hat{\mathbf{x}},$$

and

$$\mathbf{B}_{\mathrm{I}} = \frac{1}{v_1} E_{0\mathrm{I}} e^{i(k_1 z - \omega t)} \hat{\mathbf{y}}.$$

Assuming the electric field does not switch directions, the fields of the reflected wave are

$$\mathbf{E}_{\mathrm{R}} = E_{0\mathrm{R}} e^{i(-k_1 z - \omega t)} \hat{\mathbf{x}},$$

and

$$\mathbf{B}_{\mathrm{R}} = -\frac{1}{v_1} E_{0\mathrm{R}} e^{i(-k_1 z - \omega t)} \hat{\mathbf{y}}.$$

If the electric field switches direction, which depends on the relative indices of refraction, then the solution for $E_{0\mathrm{R}}$ will be negative.

The transmitted wave may be written

$$\mathbf{E}_{\mathrm{T}} = E_{0\mathrm{T}} e^{i(k_2 z - \omega t)} \hat{\mathbf{x}},$$

and

$$\mathbf{B}_{\mathrm{T}} = \frac{1}{v_2} E_{0\mathrm{T}} e^{i(k_2 z - \omega t)} \hat{\mathbf{y}}.$$

The boundary condition on parallel E gives

$$E_{0\mathrm{I}} + E_{0\mathrm{R}} = E_{0\mathrm{T}}.$$

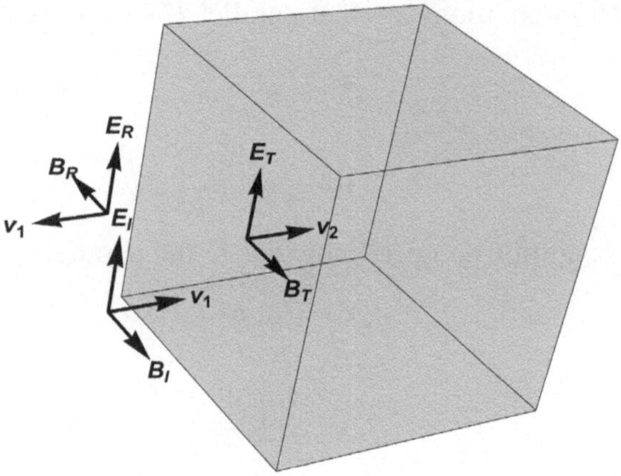

Figure 10.3 An incoming wave in medium 1 is incident at a right angle to the boundary between media 1 and 2 (shaded). Part of the wave gets reflected back into medium 1, and a portion of the wave gets transmitted into medium 2.

The boundary condition on parallel B gives

$$\frac{1}{\mu_1}\left(\frac{1}{v_1}E_{0I} - \frac{1}{v_1}E_{0R}\right) = \frac{1}{\mu_2}\frac{1}{v_2}E_{0T}.$$

The solution may be written

$$E_{0I} - E_{0R} = \beta E_{0T},$$

with

$$\beta = \frac{\mu_1 v_1}{\mu_2 v_2}.$$

Example 10.8 Solve for E_{0R} and E_{0T}.

```
In[14]:= ClearAll["Global`*"];
         Solve[E0I + E0R == E0T && E0I - E0R == β E0T, {E0R, E0T}] //
          Simplify
```

$$\text{Out[14]=}\ \left\{\left\{E_{0R} \to \frac{E_{0I} - \beta E_{0I}}{1+\beta},\ E_{0T} \to \frac{2\,E_{0I}}{1+\beta}\right\}\right\}$$

For the case $\mu_1 = \mu_2 = \mu_0$,

$$\beta = \frac{v_1}{v_2},$$

and

$$E_{0R} = \frac{v_2 - v_1}{v_2 + v_1} E_{0I} = \frac{n_1 - n_2}{n_1 + n_2} E_{0I},$$

$$E_{0T} = \frac{2v_2}{v_2 + v_1} E_{0I} = \frac{2n_1}{n_1 + n_2} E_{0I},$$

If $n_2 > n_1$, the electric field of the reflected wave switches direction.

The reflection (R) and transmission (T) coefficients are calculated from the ratios of intensities (field squared),

$$R = \frac{I_R}{I_I} = \left(\frac{E_{0R}}{E_{0I}}\right)^2,$$

and

$$T = \frac{I_R}{I_I} = \left(\frac{E_{0T}}{E_{0I}}\right)^2,$$

Example 10.9 Show that $R + T = 1$

```
In[15]:= β = μ1 v1 / (μ2 v2);
         (E0R /. s[[1]] / E0I)^2 + (ε2 v2)/(ε1 v1) (E0T /. s[[1]] / E0I)^2 /.
           {v1 → 1/Sqrt[ε1 μ1], v2 → 1/Sqrt[ε2 μ2]} // Simplify

Out[16]= 1
```

10.8 OBLIQUE INCIDENCE

The angles of incidence (θ_I) and reflection (θ_r) are defined w.r.t. the direction perpendicular to the planar boundary (Fig. 10.4). Consider now the case where the incident angle is not zero. For the first part of the calculation, the polarization (**E**) direction will be kept arbitrary.

The fields for the waves are

$$\mathbf{E}_I = \mathbf{E}_{0I} e^{i(\mathbf{k}_I \cdot \mathbf{r} - \omega t)},$$

$$\mathbf{B}_I = \frac{1}{v_1} \hat{\mathbf{k}}_I \times \mathbf{E}_I,$$

$$\mathbf{E}_R = \mathbf{E}_{0R} e^{i(\mathbf{k}_R \cdot \mathbf{r} - \omega t)},$$

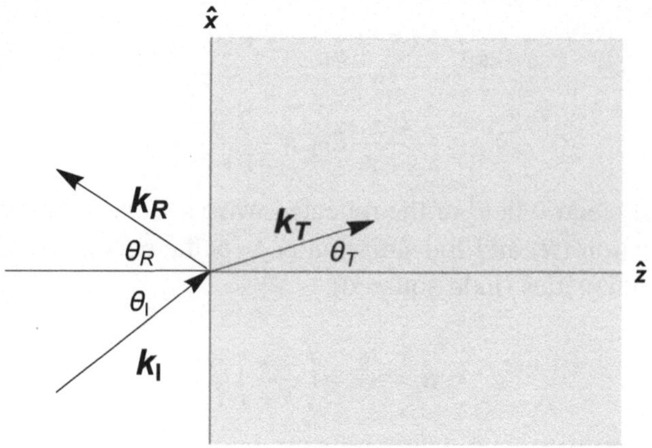

Figure 10.4 An incoming wave has an incident angle $\theta_{\rm i}$, relected angle $\theta_{\rm R}$, and transmitted angle $\theta_{\rm T}$.

$$\mathbf{B}_{\rm R} = \frac{1}{v_1}\hat{\mathbf{k}}_{\rm R} \times \mathbf{E}_{\rm R},$$

$$\mathbf{E}_{\rm T} = \mathbf{E}_{0{\rm T}} e^{i(\mathbf{k}_{\rm T}\cdot\mathbf{r}-\omega t)},$$

and

$$\mathbf{B}_{\rm T} = \frac{1}{v_2}\hat{\mathbf{k}}_{\rm T} \times \mathbf{E}_{\rm T}.$$

All three waves have the same frequency (which came from the source),

$$\omega = k_{\rm I}v_1 = k_{\rm R}v_1 = k_{\rm T}v_2.$$

The wave numbers are related by

$$k_{\rm I} = k_{\rm R} = \frac{v_2}{v_1}k_{\rm T} = \frac{n_1}{n_2}k_{\rm T}.$$

10.8.1 Relationships Between the Angles

The boundary condition at $z = 0$ requires all the exponents to match,

$$\mathbf{k}_{\rm I}\cdot\mathbf{r} = \mathbf{k}_{\rm R}\cdot\mathbf{r} = \mathbf{k}_{\rm T}\cdot\mathbf{r},$$

for all values of x and y, giving

$$xk_{{\rm I}x} + yk_{{\rm I}y} = xk_{{\rm R}x} + yk_{{\rm R}y} = xk_{{\rm T}x} + yk_{{\rm T}y}.$$

Applying this at $y = 0$ gives

$$k_{Ix} = k_{Rx} = k_{Tx},$$

and at $x = 0$ gives

$$k_{Iy} = k_{Ry} = k_{Ty},$$

If the incident wave vector is in the $x-z$ plane, then the reflected and transmitted wave vectors are also in that plane. Since $\mathbf{k}_I$ and $\mathbf{k}_R$ are in the same magnitude and have the same x and y components at $z = 0$, it follows that

$$\theta_I = \theta_R.$$

Similarly since $\mathbf{k}_I$ and $\frac{n_1}{n_2}\mathbf{k}_T$ are the same magnitude and have the same x and y components at $z = 0$, it follows that

$$k_I \sin\theta_I = k_T \sin\theta_T = \frac{n_2}{n_1} k_I.$$

This gives Snell's law of refraction,

$$n_1 \sin\theta_I = n_2 \sin\theta_T.$$

10.8.2 Choosing the Polarization

The details of the solution depend on the choice of polarization. Consider the case where the electric field is in the $x-z$ plane (Fig. 10.5).

The boundary condition on $\mathbf{E}^{\perp}$ gives

$$-\varepsilon_1 E_{0I} \sin\theta_i + \varepsilon_1 E_{0R} \sin\theta_R = -\varepsilon_2 E_{0T} \sin\theta_T.$$

The boundary condition on $\mathbf{E}^{\|}$ gives

$$E_{0I} \cos\theta_i + E_{0R} \cos\theta_R = E_{0T} \cos\theta_T.$$

The boundary condition on $\mathbf{B}^{\|}$ gives

$$\frac{1}{\mu_1\varepsilon_1} E_{0I} - \frac{1}{\mu_1\varepsilon_1} E_{0R} = \frac{1}{\mu_2\varepsilon_2} E_{0T}.$$

Define

$$\beta = \frac{\mu_1 v_1}{\mu_2 v_2},$$

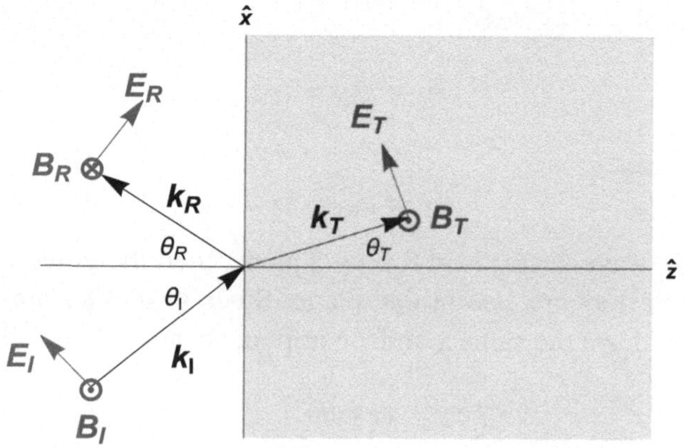

Figure 10.5 An incoming wave is polarized in the $x-y$ plane.

and

$$\alpha = \frac{\cos\theta_{\mathrm{T}}}{\cos\theta_{\mathrm{I}}},$$

to give

$$E_{0\mathrm{I}} - E_{0\mathrm{R}} = \beta E_{0\mathrm{T}},$$

and

$$E_{0\mathrm{I}} + E_{0\mathrm{R}} = \alpha E_{0\mathrm{T}},$$

Example 10.10 Solve for $E_{0\mathrm{R}}$ and $E_{0\mathrm{T}}$.

```
In[17]:= ClearAll["Global`*"];
         Solve[E0I + E0R == α E0T && E0I - E0R == β E0T, {E0R, E0T}] //
          Simplify
```

Out[17]= $\left\{\left\{E_{0R} \to \frac{(\alpha - \beta)\, E_{0I}}{\alpha + \beta},\ E_{0T} \to \frac{2\, E_{0I}}{\alpha + \beta}\right\}\right\}$

These equations are known as the Fresnel equations for the case of polarization in the plane of incidence. The result agrees with normal incidence when $\alpha = 1$. At an incident angle of $\pi/2$, $\alpha \to \infty$ and the wave is totally reflected.

10.8.3 Brewster Angle

Examination of the reflected electric field shows that there is an incident angle for which there is no reflected wave, occurring at

$$\alpha = \beta = \frac{\mu_1 v_1}{\mu_2 v_2} = \frac{n_2}{n_1}.$$

From the definition of α together with Snell's law,

$$\alpha = \frac{\cos\theta_T}{\cos\theta_I} = \frac{\sqrt{1-\sin^2\theta_T}}{\cos\theta_I} = \frac{\sqrt{1-\frac{n_1}{n_2}\sin^2\theta_I}}{\cos\theta_I}.$$

Thus,

$$\alpha\cos\theta_I = \sqrt{1-\frac{n_1}{n_1}\sin^2\theta_I}.$$

Squaring and using $\alpha = \beta = n_2/n_1$,

$$\beta^2(1-\sin^2\theta_I) = 1 - \frac{\sin^2\theta_I}{\beta^2},$$

Example 10.11 Solve for $\sin^2\theta_I$.

```
In[18]:= Clear[β]; Solve[β² (1 - x) == 1 - x/β², x]
```

```
Out[18]= {{x → β²/(1 + β²)}}
```

The solution is

$$\sin^2\theta_I = \frac{\beta^2}{1+\beta^2}.$$

Labeling the incident angle as the Brewster angle θ_B,

$$\cos^2\theta_B = 1-\sin^2\theta_B = 1-\frac{\beta^2}{1+\beta^2} = \frac{1}{1+\beta^2},$$

and

$$\tan\theta_B = \beta = \frac{n_2}{n_1}.$$

Figure 10.6 shows a plot of E_{0R}/E_{0I} and E_{0T}/E_{0I} as a function of incident angle. The Brewster angle is where the reflected wave crosses the x-axis.

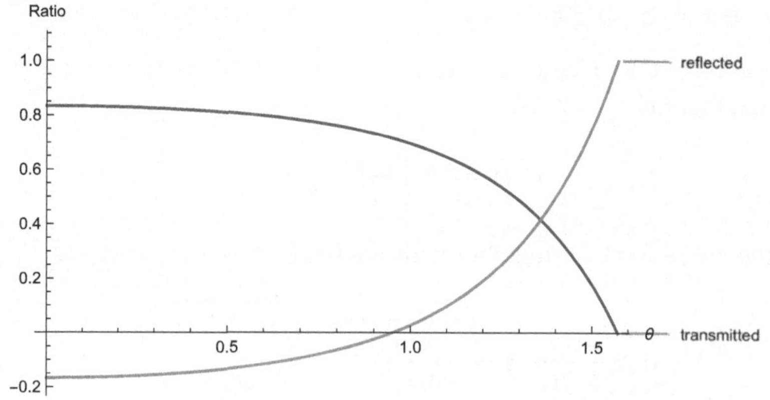

Figure 10.6 The ratios E_{0R}/E_{0I} and E_{0T}/E_{0I} are plotted as a function of incident angle

The reflected intensity relative to the incident intensity is given by the square of the electric field ratio,

$$R = \left(\frac{E_{0R}}{E_{0I}}\right)^2 = \left(\frac{\alpha-\beta}{\alpha+\beta}\right)^2.$$

The transmitted relative intensity has the additional factor

$$T = \frac{\varepsilon_2 v_2 \cos\theta_T}{\varepsilon_1 v_1 \cos\theta_I}\left(\frac{E_{0T}}{E_{0I}}\right)^2 = \alpha\beta\left(\frac{2}{\alpha+\beta}\right)^2.$$

Figure 10.7 shows a plot of $(E_{0R}/E_{0I})^2$ and $(E_{0T}/E_{0I})^2$ as a function of incident angle.

Example 10.12 Show that $R+T=1$.

```
In[19]:= ClearAll["Global`*"];
         Simplify[α β (2 / (α + β))^2 + ((α - β) / (α + β))^2]

Out[19]= 1
```

Example 10.13 Find the incident angle at which the reflected intensity from air off a diamond ($n = 2.417$) is 1%.

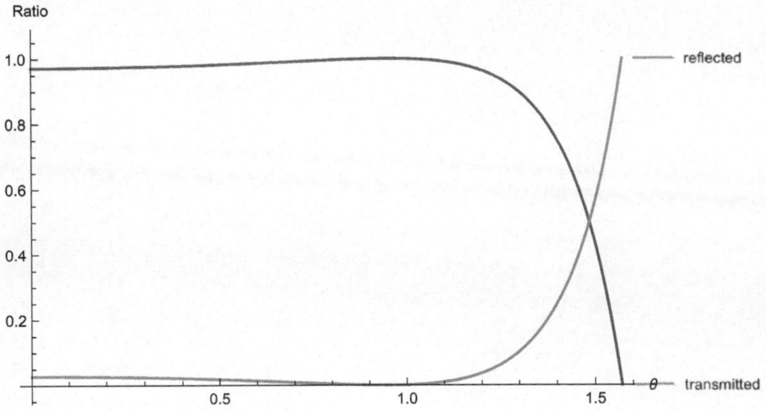

Figure 10.7 The ratios $(E_{0R}/E_{0I})^2$ and $(E_{0T}/E_{0I})^2$ are plotted as a function of incident angle

```
In[20]:= n1 = 1; n2 = 2.417;
```

$$\alpha = \frac{\sqrt{1 - \left(\frac{n_1}{n_2}\,\mathrm{Sin}[\theta_{0\,I}]\right)^2}}{\mathrm{Cos}[\theta_{0\,I}]};\ \beta = \frac{\mu_1\, n_2}{\mu_1\, n_1};$$

$$\mathrm{Solve}\left[\left(\frac{\alpha - \beta}{\alpha + \beta}\right)^2 == \frac{1}{100},\ \theta_{0\,I}\right]$$

$$\text{Out[22]=}\ \{\{\theta_{0\,I} \to -1.25429\},\ \{\theta_{0\,I} \to -1.08061\},\ \{\theta_{0\,I} \to 1.08061\},\ \{\theta_{0\,I} \to 1.25429\}\}$$

There are seen to be two unique angles as expected, corresponding to the two possible values of the reflected field (see Fig.10.7).

CHAPTER 11

Fields in Moving Frames of Reference

11.1 BASICS OF SPECIAL RELATIVITY

The two basic postulates of special relativity are that
1) the laws of physics are identical in all inertial frames of reference, and
2) the speed of light in vacuum (c) is the same for all observers.
This leads to the phenomena of time dilation, that time intervals are longer on moving clocks,

$$\Delta t' = \gamma \Delta t,$$

where γ is defined by the relative speed v of the moving frame

$$\gamma = \frac{1}{\sqrt{1-\beta^2}},$$

and $\beta = v/c$. The postulates of special relativity also lead to the phenomena of length contraction,

$$L' = \frac{L}{\gamma},$$

where lengths of moving objects are measured to be shorter.

One should not get hung upon which variable is called "prime," as it does not matter. The moving clock has a longer time interval and the moving stick is shorter.

Time dilation and length contraction go together hand-in-hand. For example, a cosmic muon produced at the top of the atmosphere is, in the earth's frame, a moving clock that lives longer before spontaneous decay compared to the rest frame of the muon where the surface of the earth is moving toward it and is length-contracted. In the first frame, the lifetime is longer by gamma, and in the second frame the distance traveled is shorter by gamma. As analyzed in each frame, the muon reaches the surface of the earth before decaying, and in both frames the relative speed of the muon with respect to the surface of the earth is $v = \Delta L/\Delta t$.

11.2 FOUR-VECTORS

A four-vector is a quantity that transforms in a special way when going from one frame to another with relative speed v. The four-vector X is written

$$\mathrm{X} = (x_0, x_1, x_2, x_3),$$

where a zeroth piece has been added to an ordinary vector $\mathbf{x}$. The length squared of the four-vector is defined to be

$$\mathrm{X}\cdot\mathrm{X} = x_0^2 - \mathbf{x}\cdot\mathbf{x} = x_0^2 - (x_1^2 + x_2^2 + x_3^2).$$

The product of two four-vectors A and B is

$$\mathrm{A}\cdot\mathrm{B} = A_0B_0 - (A_1B_1 + A_2B_2 + A_3B_3).$$

Unfortunately, not everybody uses the same sign convention. The most common alternative is to introduce a minus sign such that the length becomes $\mathbf{x}\cdot\mathbf{x} - x_0^2$. This convention is generally used in astrophysics, while particle physics generally uses the sign convention introduced here.

The convention is to use a Greek letter to specify a component of a four-vector compared to the use of a Roman letter for an ordinary vector. The sign is accounted for by defining "lower" four-vectors

$$X_\mu = (x_0, -x_1, -x_2, -x_3)$$

and "upper" four-vectors

$$X^\mu = (x_0, x_1, x_2, x_3).$$

Then the square (or product) is always a lower times an upper,

$$\sum_{\mu=0}^{3} X_\mu X^\mu = x_0^2 - (x_1^2 + x_2^2 + x_3^2).$$

A simple tensor g may be formed to transform between lower and upper four-vectors (in either direction),

$$g = \begin{pmatrix} 1 & 0 & 0 & 0 \\ 0 & -1 & 0 & 0 \\ 0 & 0 & -1 & 0 \\ 0 & 0 & 0 & -1 \end{pmatrix},$$

such that going from upper (lower) to lower (upper) is a simple matrix multiplication,

$$\begin{pmatrix} x_0 \\ x_1 \\ x_2 \\ x_3 \end{pmatrix} = g \begin{pmatrix} x_0 \\ -x_1 \\ -x_2 \\ -x_3 \end{pmatrix},$$

and

$$\begin{pmatrix} x_0 \\ -x_1 \\ -x_2 \\ -x_3 \end{pmatrix} = g \begin{pmatrix} x_0 \\ x_1 \\ x_2 \\ x_3 \end{pmatrix}.$$

Example 11.1 Construct the tensor g and transform a lower four-vector to an upper four-vector.

```
In[1]:= g = {{1, 0, 0, 0}, {0, -1, 0, 0}, {0, 0, -1, 0},
            {0, 0, 0, -1}};
        MatrixForm[g]
        X = {x₀, x₁, x₂, x₃}
        g.X
```

Out[1]//MatrixForm=

$$\begin{pmatrix} 1 & 0 & 0 & 0 \\ 0 & -1 & 0 & 0 \\ 0 & 0 & -1 & 0 \\ 0 & 0 & 0 & -1 \end{pmatrix}$$

Out[2]= $\{x_0, x_1, x_2, x_3\}$

Out[3]= $\{x_0, -x_1, -x_2, -x_3\}$

The length of a four-vector is then obtained from XgX where the X could be either an upper or a lower.

Example 11.2 Calculate the length of a four-vector.

```
In[4]:= √X.g.X
```

Out[4]= $\sqrt{x_0^2 - x_1^2 - x_2^2 - x_3^2}$

11.3 LORENTZ TRANSFORMATION

In transforming between frames with relative speed v, the Lorentz transformation Λ matrix preserves the length of the four-vector. The form of Λ for relative motion in the x direction with $v = \beta c$ is

$$\Lambda = \begin{pmatrix} \gamma & \beta\gamma & 0 & 0 \\ \beta\gamma & 1 & 0 & 0 \\ 0 & 0 & 1 & 0 \\ 0 & 0 & 0 & 1 \end{pmatrix}.$$

To transform in the opposite direction, switch the sign of β.

Example 11.3 Construct the Lorentz transformation for relative motion in the x direction with relative speed $v = \beta c$.

```
In[5]:= Λ = {{γ, βγ, 0, 0}, {βγ, γ, 0, 0}, {0, 0, 1, 0},
          {0, 0, 0, 1}};
        MatrixForm[Λ]

Out[5]//MatrixForm=
  ( γ   βγ  0  0 )
  ( βγ  γ   0  0 )
  ( 0   0   1  0 )
  ( 0   0   0  1 )
```

Example 11.4 Verify that the Lorentz transformation does not change the length of a four-vector.

```
In[6]:= X = {x₀, x₁, x₂, x₃};
        X.g.X == FullSimplify[(Λ.X).g.(Λ.X)] /. -1 + β^2 → -γ^-2

Out[6]= True
```

For a four-vector to carry any physical meaning, there must be a reason that its length is the same in all frames of reference. For example, (ct, x, y, z) makes a four-vector because the speed of light is invariant (the same in all frames of reference). Other important examples of four-vectors include energy-momentum $(E, \mathbf{p}c)$ because mass is invariant, charge-current density $(\rho c, \mathrm{J})$ because charge is invariant, and scalar-vector potential $(V/c, \mathbf{A})$ which follows from the charge-current density four-vector. The electric and magnetic fields are not part of four-vectors. They do not have simple transformations.

11.3.1 Time Dilation

If times t_1 and t_2 are measured at the same position (x, y, z) in frame S, then in frame S' that is moving in the x direction with relative speed v, the corresponding time interval is given by the Lorentz transformation,

$$c(t_2' - t_1') = (\gamma c t_2 + \beta\gamma x) - (\gamma c t_1 + \beta\gamma x) = \gamma c(t_2 - t_1),$$

or

$$t_2' - t_1' = \gamma(t_2 - t_1).$$

The time integral is longer in frame S'.

Example 11.5 Transform the time interval.

```
In[7]:= X₁ = {c t₁, x, y, z}; X₂ = {c t₂, x, y, z};
        (Λ.X₂ - Λ.X₁).{1, 0, 0, 0}

Out[8]= -c γ t₁ + c γ t₂
```

11.3.2 Length Contraction

Consider a stationary clock that measures the time interval Δt for a stick moving with speed v to pass (Fig. 11.1).

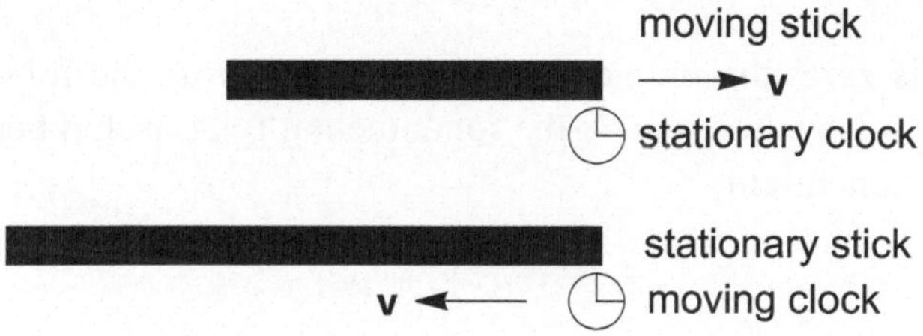

Figure 11.1 The length of a stick is measured with stationary and moving clocks.

The speed of the stick is

$$v = \frac{\Delta x}{\Delta t},$$

where Δx is the length of the stick measured in that frame. Now perform the same measurement in a frame where the stick is at rest and the clock is moving with speed v. Now

$$v = \frac{\Delta x'}{\Delta t'} = \frac{\Delta x'}{\gamma \Delta t},$$

where the last step is from time dilation. Equating the two expressions for v gives

$$\Delta x = \frac{\Delta x'}{\gamma},$$

and the stick is measured to be shorter in the frame where it is moving.

11.4 ENERGY-MOMENTUM FOUR-VECTOR

Mass is a Lorentz invariant. It is calculated from the length of the energy-momentum four-vector. The mass energy (energy that is stored as mass) is

$$E_0 = mc^2.$$

The total energy E is mass energy plus kinetic energy K

$$E = E_0 + K.$$

The momentum, provided the speed is not zero, is

$$p = \frac{m\mathbf{v}}{\sqrt{1-(v/c)^2}} = \gamma m\mathbf{v}.$$

If the mass is zero, the momentum is calculated from the following equation which always is true and is the fundamental relationship between mass, energy, and momentum,

$$E^2 = (mc^2)^2 + (pc)^2.$$

Thus,

$$mc^2 = \sqrt{E^2 - (pc)^2},$$

and $(E, \mathbf{p}c)$ makes a four-vector with length equal to mc^2.

11.4.1 Speed

A mass m at rest has a four-vector

$$P = (mc^2, 0).$$

Transforming it to frame where it has a speed, its new four-vector is

$$P' = (\gamma mc^2, \beta\gamma mc^2).$$

Example 11.6 Transform the four-vector for a mass m at rest to a frame where it has speed v.

```
In[9]:= P = {m c^2, 0, 0, 0};
        MatrixForm [Λ.P]
Out[10]//MatrixForm=
```

$$\begin{pmatrix} c^2\, m\, \gamma \\ c^2\, m\, \beta\, \gamma \\ 0 \\ 0 \end{pmatrix}$$

It is seen from the transformed four-vector that

$$\beta = \frac{v}{c} = \frac{pc}{E}.$$

If m is not zero, the total energy is

$$E = \sqrt{(mc^2)^2 + (pc)^2},$$

and a convenient expression for particle velocity is

$$\mathbf{v} = \frac{\mathbf{p}c}{E}c,$$

provided the mass is not zero. If the mass is zero, then $pc = E$ and the particle speed is c.

11.4.2 Total Energy

The expression for total energy is

$$E = \sqrt{(mc^2)^2 + (pc)^2} = mc^2\sqrt{1 + \frac{(pc)^2}{(mc^2)^2}}.$$

Since $mc^2 = E/\gamma$, $pc/E = \beta$, and $\beta^2 = 1 - 1/\gamma^2$,

$$E = mc^2\sqrt{1 + \gamma^2\beta^2} = \gamma mc^2$$

as long as m is not zero.

11.4.3 Kinetic Energy

The exact expression for kinetic energy is

$$K = E - mc^2 = \sqrt{(mc^2)^2 + (pc)^2} - mc^2.$$

In the nonrelativistic limit, the kinetic energy is seen to reduce to the familiar form

$$K = \frac{p^2}{2m} = \frac{1}{2}mv^2.$$

Example 11.7 Expand the square-root term in the exact expression for kinetic energy to get the nonrelativistic approximation.

```
In[11]:= $Assumptions = {m > 0, p > 0, c > 0};
         K = √((m c²)² + (p c)²) - m c²;
         Series[K, {p, 0, 2}]

Out[11]= p²/(2 m) + O[p]³
```

Consider an electron with momentum 1 MeV/c.

Example 11.8 Calculate the total energy.

```
In[12]:= E = √((electron PARTICLE [mass] c²)² + (1 MeV)²)

Out[12]= 1122.9959585 keV
```

Example 11.9 Calculate the kinetic energy.

```
In[13]:= UnitConvert[E - electron PARTICLE [mass] c², MeV]

Out[13]= 0.6119970124 MeV
```

Example 11.10 Calculate β.

```
In[14]:= 1 MeV / E

Out[14]= 0.8904751548
```

Example 11.11 Calculate γ.

```
In[15]:= UnitConvert[E/ (electron PARTICLE [mass] c²)]

Out[15]= 2.197648287
```

11.4.4 Velocity Addition

The traditional derivation of the "addition" of velocities using the Lorentz transformation of dx/dt is usually not useful. It is far easier and more intuitive to transform the momentum. Consider two particles that are moving toward each other with four-vectors P_1 and P_2 (Fig. 11.2).

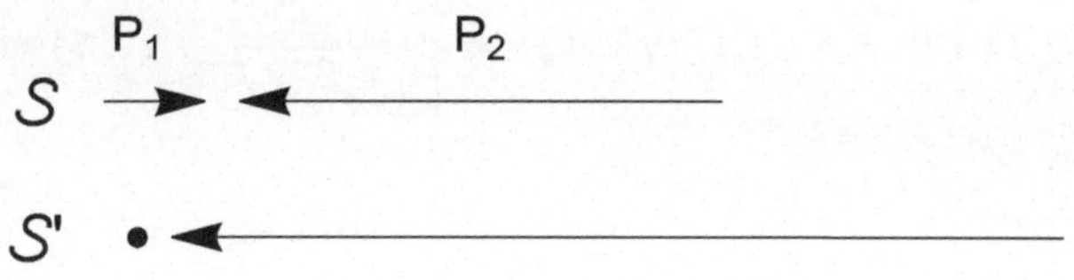

Figure 11.2 Two particles with four-vectors P_1 and P_2 in frame $\mathcal{S}$ approach each other. In frame $\mathcal{S}'$ one of the particles is at rest and the velocity of the other particle is the relative velocity of the two particles.

Let γ and β correspond to the transform that brings particle 1 to rest. Then $c\beta$ is the speed of particle 1 in frame $\mathcal{S}$,

$$\beta = \frac{v_1}{c}.$$

The speed of particle 2 in $\mathcal{S}'$ is given by

$$\frac{v_2}{c} = \frac{p_2'c}{E_2'} = \frac{\beta\gamma E_2 + \gamma p_2 c}{\gamma E_2 + \beta\gamma p_2 c} = \frac{\beta + p_2 c/E_2}{1 + p_2 c/E_2},$$

or

$$v_2' = \frac{v_1 + v_2}{1 + v_1 v_2/c^2}.$$

Speed is not a good variable in special relativity and most of the time it is more intuitive to express the speed in terms of gamma. For this case

$$\frac{E_2'}{mc^2} = \frac{\gamma E_2 + \beta\gamma p_2 c}{mc^2} = \gamma\gamma_2 + \gamma\beta\gamma_2\beta_2.$$

Thus, if two particles having γ_1 and γ_2 approach each other, their relative γ is

$$\gamma = \gamma_1\gamma_2(1 + \beta_1\beta_2).$$

Example 11.12 Show that $\gamma_1\gamma_2(1 + \beta_1\beta_2) = 1/\sqrt{1-(v/c)^2}$ where $v = (v_1 + v_2)(1 + v_1 v_2/c^2)$.

```
In[16]:= $Assumptions = {0 < β1 < 1, 0 < β2 < 1}; γ1 = 1/Sqrt[1 - β1^2];
         γ2 = 1/Sqrt[1 - β2^2]; v = (c β1 + c β2)/(1 + β1 β2);

         (γ1 γ2 (1 + β1 β2) // Simplify) == (1/Sqrt[1 - (v/c)^2] // Simplify)

Out[18]= True
```

Example 11.13 Two particles approach each other each traveling at 0.9 c. Calculate their relative speed and the corresponding γ.

```
In[19]:= β1 = .9; β2 = .9;
         v
         1/Sqrt[1 - (v/c)^2]

Out[20]= 0.994475 c

Out[21]= 9.52632
```

11.5 EXAMPLES OF FIELDS IN MOVING FRAMES

11.5.1 Moving Capacitor

Consider a capacitor at rest in frame S with charge density σ_0 (Fig. 11.3). The electric field is

$$E = \frac{\sigma_0}{\varepsilon_0}.$$

Viewed in a frame S' where the capacitor is moving with speed v in a direction parallel to the plates (Fig. 11.4), the charge is Lorentz-contracted and the electric field is

$$E' = \frac{\sigma}{\varepsilon_0} = \frac{\gamma\sigma_0}{\varepsilon_0}.$$

In a frame where the capacitor is moving in a direction perpendicular to the plates (Fig. 11.5), the field is unchanged.

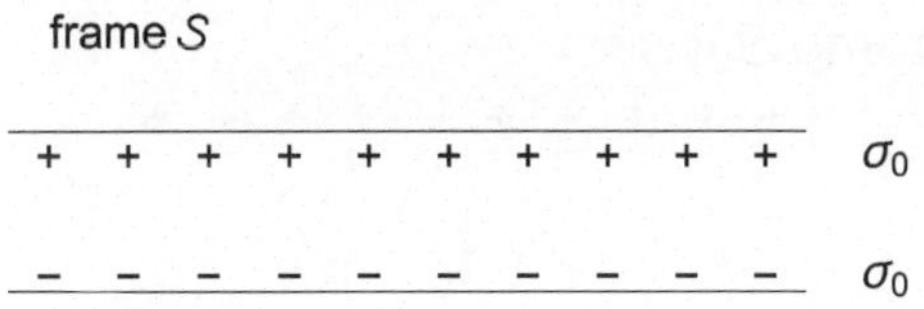

Figure 11.3 A capacitor is at rest.

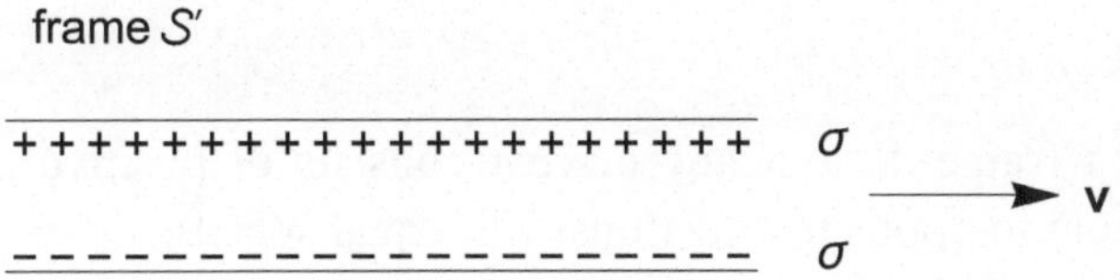

Figure 11.4 A capacitor is moving with speed v in a direction parallel to its plates. The spacing of the charge is Lorentz-contracted along the direction of motion.

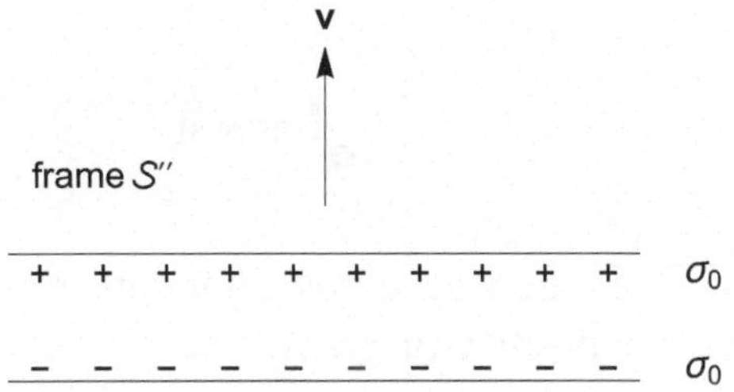

Figure 11.5 A capacitor is moving in a direction perpendicular to the plates. The charge distribution is unchanged.

11.5.2 Steady Current

Consider long lines of both positive and negative charges moving in opposite directions to form a steady current (Fig. 11.6). For charge densities $\pm\lambda$, the current is $I = 2\lambda v$, and the Lorentz force on a nearby charge q moving with speed u is

$$F = quB = qu\frac{\mu_0 I}{2\pi r} = \frac{\mu_0 q\lambda uv}{\pi r}.$$

Th magnetic force is attractive for positive q. There is no electric force on q.

Now consider the frame S' where q is at rest (Fig. 11.7). The positive charge is moving faster to the left and is Lorentz-contracted, while the negative charge is moving slower to the right and is expanded. In this frame there are both electric and magnetic forces on q.

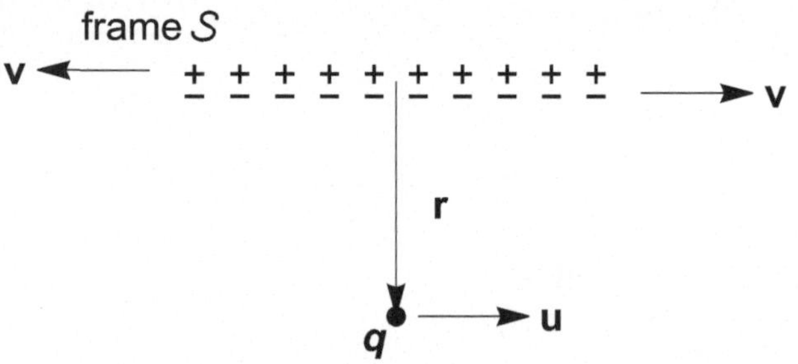

Figure 11.6 In frame S, a steady current consists of positive and negative charges moving in opposite directions with equal speeds.

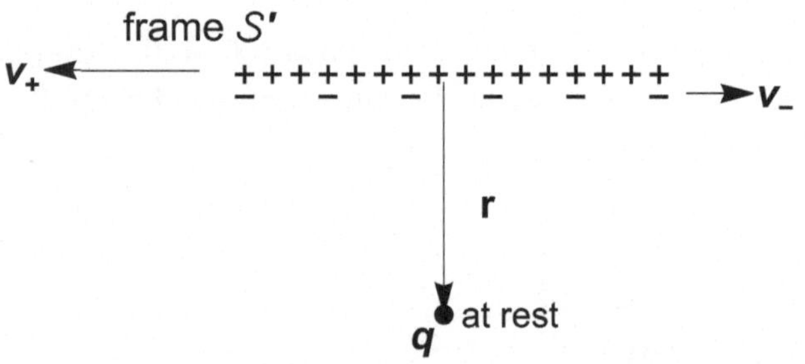

Figure 11.7 In a frame S', moving in the same direction as the motion of the negative charges, the positive (negative) charges have a larger (smaller) speed.

There are two velocity additions (Sect. 11.4.4) to solve, adding u to v and subtracting u from v, then what is really needed are the γ factors that correspond to these resulting speeds. Let γ correspond to the original speed v of the charges,

$$\gamma = \frac{1}{\sqrt{1-(v/c)^2}}.$$

Let γ_+ (γ_-) correspond to the motion of the positive (negative) charges,

$$\gamma_+ = \frac{1}{\sqrt{1-(v_+/c)^2}},$$

and

$$\gamma_- = \frac{1}{\sqrt{1-(v_-/c)^2}},$$

The speed v_+ (v_-) is just the relative speed of the point charge and the line of charges moving in the opposite (same) direction. The expressions for γ_+ and γ_- are just energy to mass energy ratios of a charge with speed $u = \frac{pc}{E}$ boosted by γ,

$$\gamma_+ = \frac{\gamma E + \beta\gamma pc}{\sqrt{E^2 - (pc)^2}} = \frac{1 + \frac{uc}{c^2}}{\sqrt{E^2 - (pc)^2}},$$

and

$$\gamma_- = \frac{\gamma E - \beta\gamma pc}{\sqrt{E^2 - (pc)^2}} = \frac{1 - \frac{uc}{c^2}}{\sqrt{E^2 - (pc)^2}},$$

The gamma factors give the Lorentz contraction of the charge in different frames. Let λ_0 be the "proper" charge density, the charge density in its rest frame. The charge density in frame S is

$$\lambda = \gamma\lambda_0.$$

The charge densities in frame S' are

$$\lambda_+ = \gamma_+\lambda_0,$$

and

$$\lambda_- = \gamma_-\lambda_0.$$

The total charge density λ_{tot} in frame S' is

$$\lambda_{tot} = \lambda_+ + \lambda_- = \lambda_0(\gamma_+ - \gamma_-) = \lambda_0\gamma\frac{2uv/c^2}{\sqrt{1-(u/c)^2}}.$$

The electric field in frame S' is

$$E' = \frac{\lambda_{tot}}{2\pi\varepsilon_0 r} = \frac{\lambda_0\gamma}{\pi\varepsilon_0 r}\frac{uv/c^2}{\sqrt{1-(u/c)^2}} = \frac{\lambda}{\pi\varepsilon_0 r}\frac{uv/c^2}{\sqrt{1-(u/c)^2}}.$$

Comparing to the magnetic field in frame S,

$$B = \frac{\mu_0\lambda v}{\pi r} = \frac{\lambda v}{\pi\varepsilon_0 c^2 r} = \gamma_u u E',$$

where

$$\gamma_u = \frac{1}{\sqrt{1-(u/c)^2}},$$

the gamma that corresponds to the speed of the charge in frame S. In frame S' there is no magnetic force and the electric force on q is

$$F' = qE' = \gamma_u\frac{q\lambda uv}{\pi\varepsilon_0 c^2 r}.$$

In frame $\mathcal{S}$ there is no electric force and the magnetic force on q is

$$F = \frac{\mu_0 q\lambda uv}{\pi r} = \frac{q\lambda uv}{\pi\varepsilon_0 c^2 r} = \frac{F'}{\gamma_u}.$$

The magnetic force is smaller by a factor of γ_u. Examination of the momentum transfer shows that in frame $\mathcal{S}$

$$\Delta p = F\Delta t,$$

while in frame $\mathcal{S}'$,

$$\Delta p' = F'\Delta t' = F\gamma_u\Delta t' = F\Delta t$$

due to time dilation. The momentum transfer is calculated to be identical in each frame although one force is electric, larger, and acts for shorter time while the other is magnetic, smaller, and acts for a longer time.

Example 11.14 Choose the charge to be a proton, $v = c/2$, $u = c/4$, $\lambda = 10^{-6}$ C/m, and $r = 1$ m. Calculate the magnetic force in frame $\mathcal{S}$ and the electric force in frame $\mathcal{S}'$.

```
In[22]:= q = proton PARTICLE [ electric charge ]; v = c / 2; u = c / 4;
         λ = 10^-6 C / m;
         r = 1 m; γu = 1 / Sqrt[1 - (u / c)^2];
         B = N[UnitConvert[μ0 2 λ v / (2 π r), T], 2];
         FB = UnitConvert[q u B, eV/m]
         FE = FB γu

Out[24]= 4.5×10^3 eV/m

Out[25]= 4.6×10^3 eV/m
```

11.6 POINT CHARGE WITH CONSTANT VELOCITY

Take the velocity to be in the x direction. The charge will be analyzed in two frames. In frame $\mathcal{S}$ the charge moves with a speed v in the x direction, and in frame $\mathcal{S}'$ the charge is at rest (Fig. 11.8).

In frame $\mathcal{S}'$, the distance to the charge is

$$R' = \sqrt{x'^2 + y'^2 + z'^2},$$

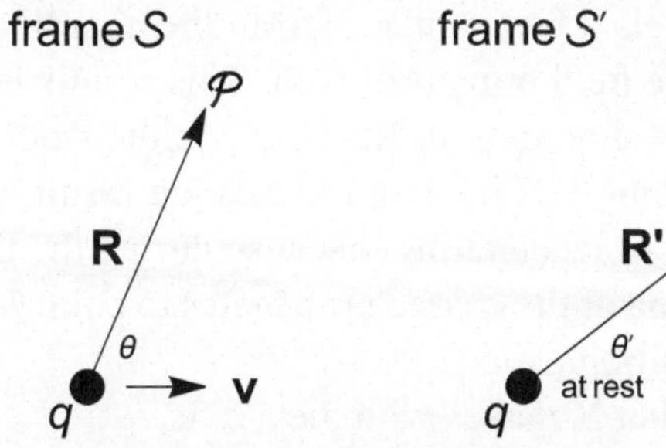

Figure 11.8 In frame S a point charge moves with velocity v in the x direction. In frame S' the charge is at rest.

and the electric field components are

$$E'_x = \frac{qx'}{4\pi\varepsilon_0(x'^2 + y'^2 + z'^2)^{3/2}},$$

$$E'_y = \frac{qy'}{4\pi\varepsilon_0(x'^2 + y'^2 + z'^2)^{3/2}},$$

and

$$E'_z = \frac{qz'}{4\pi\varepsilon_0(x'^2 + y'^2 + z'^2)^{3/2}}.$$

In frame S,

$$R = \sqrt{x^2 + y^2 + z^2}.$$

Transforming the coordinates,

$$x' = \gamma x \qquad y' = y \qquad z' = z.$$

(Notice that x is smaller than x' because the charge is moving in frame S.)

Transforming the field, the x and y components of the field in S get a γ factor (like the capacitor),

$$E_x = \frac{q\gamma x}{4\pi\varepsilon_0[(\gamma x)^2 + y^2 + z^2]^{3/2}},$$

$$E_y = \frac{\gamma q y}{4\pi\varepsilon_0[(\gamma x)^2 + y^2 + z^2]^{3/2}},$$

and

$$E_z = \frac{\gamma q z}{4\pi\varepsilon_0[(\gamma x)^2 + y^2 + z^2]^{3/2}}.$$

Thus, each coordinate gets a factor of γ, x from the coordinate transformation, and y and z from the field transformation. This results in an electric field that points in the same direction as **R**, which points from the present position of the charge to point $\mathcal{P}$. This is a remarkable result which is not true in general if the charge is accelerating, because during the time that the information about the position of the charge propagates to point $\mathcal{P}$, the charge has moved to a different position.

In terms of the angle θ that **R** makes with the x axis,

$$\mathbf{E} = \frac{\gamma q \mathbf{R}}{4\pi\varepsilon_0[(\gamma R\cos\theta)^2 + (R\sin\theta)^2]^{3/2}}.$$

Example 11.15 Simplify the expression for **E** in terms of $\beta = v/c$.

```
In[26]:= $Assumptions = 0 < β < 1;
              γ q
         ─────────────────────────────────── /.
         ((γ R Cos[θ])² + (R Sin[θ])²)^(3/2)

         {γ → 1/Sqrt[1 - β²], Cos[θ]² → 1 - Sin[θ]²} // Simplify

Out[26]=      (-1 + β²) (-1 e)
         ─────────────────────────────
         (-R² (-1 + β² Sin[θ]²))^(3/2)
```

The result of Ex. 11.17 gives

$$\mathbf{E} = \frac{(1-\beta^2)q\mathbf{R}}{4\pi\varepsilon_0(1-\beta^2\sin^2\theta)^{3/2}R^3}.$$

The magnetic field for the moving point charge is calculated in Sect. 11.8 by transformation of the fields.

11.7 TRANSFORMATION OF THE FIELDS

There are six field transformations. One of them has been calculated from the capacitor example (Sect. 11.5.1),

$$E'_x = E_x.$$

Consider a solenoid oriented with its axis along the x direction. In the rest frame of the solenoid, the magnetic field is

$$B_x = \mu_0 n I.$$

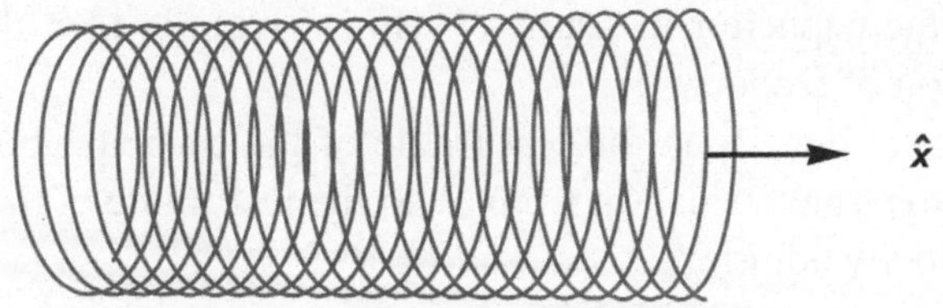

Figure 11.9 A solenoid is oriented with its axis along the x direction.

Viewed in a frame where the solenoid is moving, the number of turns per length is multiplied by γ due to length contraction, but the current is reduced by the same factor of γ due to time dilation. Therefore, the field is unchanged,

$$B'_x = \mu_0(\gamma n)(\frac{I}{\gamma}) = B_x.$$

The capacitor may be used to the other four transformations, but one needs to transform a frame that contains both electric and magnetic fields. This can be accomplished by starting with a moving capacitor (velocity u) in frame S and then transforming it to frame S' where it has a different velocity (v).

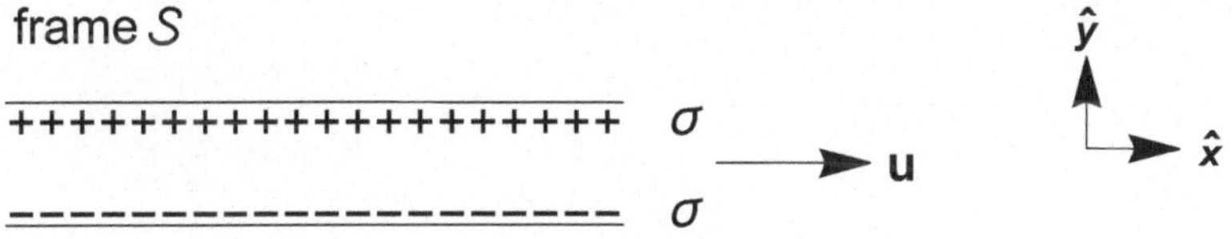

Figure 11.10 A capacitor is moving in frame S.

In frame S, the charge density is σ and the electric field is

$$E_y = \frac{\sigma}{\varepsilon_0}.$$

The currents on the plates are $\pm\sigma u$ and the magnetic field is

$$B_z = -\mu_0\sigma u.$$

In the frame S', the capacitor moves with a speed v relative to frame S. Let γ correspond to the speed v,

$$\gamma = \frac{1}{\sqrt{1-(v/c)^2}}.$$

The speed of the capacitor in frame S' (Fig. 11.11) is then the relativistic velocity addition of u and v.

Let the charge density in the rest frame of the capacitor be σ_0. Let (γ^*) correspond to the transformation from S' to the rest frame of the capacitor. It is given by velocity addition,

$$v^* = \left(1 + \frac{uv}{c^2}\right),$$

and

$$\gamma^* = \gamma\gamma_u\left(1 + \frac{uv}{c^2}\right).$$

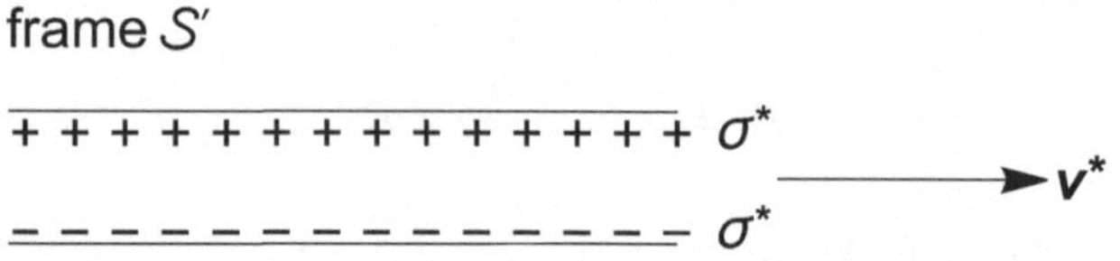

Figure 11.11 The moving capacitor is viewed in frame S' where it has a velocity v^* resulting from a transformation of frame S with relative speed v.

In frame S',

$$\sigma^* = \gamma^*\sigma_0,$$

$$E'_y = \frac{\sigma^*}{\varepsilon_0} = \frac{\gamma^*\sigma_0}{\varepsilon_0} = \frac{\gamma^*\sigma}{\gamma_u\varepsilon_0} = \gamma\left(1 + \frac{uv}{c^2}\right)\frac{\sigma}{\varepsilon_0},$$

and

$$B'_z = -\mu_0\sigma^* v^* = -\mu_0\frac{\gamma^*\sigma}{\gamma_u}\frac{u+v}{\left(1+\frac{uv}{c^2}\right)} = -\gamma\mu_0\sigma(u+v).$$

The transformations are given by comparing with the fields in frame S,

$$E'_y = \gamma(E_y - vB_z),$$

and

$$B'_z = \gamma\left(B_z - \frac{v}{c^2}E_y\right).$$

Transformations of E'_z and B'_y are obtained by exactly the same technique with the capacitor oriented in the $x-z$ plane,

$$E'z = \gamma(E_z + vB_y),$$

and

$$B'_y = \gamma\left(B_y + \frac{v}{c^2}E_z\right).$$

11.8 MAGNETIC FIELD OF A MOVING POINT CHARGE

Consider again the moving point charge of Fig. 11.8. In frame S', the magnetic field is zero. From the transformation (Sect. 11.7), the magnetic field in S is

$$B_x = 0,$$

$$B_y = \gamma\left(\frac{v}{c^2}E'_z\right),$$

and

$$B_z = \gamma\left(-\frac{v}{c^2}E'_y\right),$$

The magnetic field may be written

$$\mathbf{B} = \frac{1}{c^2}\mathbf{v}\times\mathbf{E}.$$

Thus,

$$\mathbf{B} = \frac{[1-\beta^2]q\mathbf{v}\times\mathbf{R}}{4\pi\varepsilon_0 c[1-\beta^2\sin^2\theta]^{3/2}R^3}.$$

For $\beta << 1$,

$$\mathbf{B} = \frac{q\mathbf{v}\times\mathbf{R}}{4\pi\varepsilon_0 c^2 R^3} = \frac{\mu_0 q\mathbf{v}\times\mathbf{R}}{4\pi R^3}.$$

This is what you would get from the Biot-Savart law if the moving charge was treated like a steady current. The moving charge is not a steady current and the Biot-Savart law does not hold, but as seen above it does give an approximate answer that holds true for a non-relativistic charge.

Example 11.16 Calculate the magnetic field directly at a position (1,2,3) μm, from a proton moving with 0.9 c in the $-x$ direction.

```
In[27]:= x = 1×10^-6 m; y = 2×10^-6 m; z = 3×10^-6 m; β = .9;

γ = 1/Sqrt[1 - β^2]; R = {x/γ, y, z} ;

v = -{.9 c , 0 c , 0 c };

UnitConvert[(1  β^2) e v×R / (4 π ε0 c^2 (1 - β^2 (y^2+z^2)/R.R)^(3/2) (R.R)^(3/2)), T]

Out[30]= {0. T, 5.67971×10^-7 T, -3.78647×10^-7 T}
```

Example 11.17 Calculate the magnetic field for the proton in Ex. 11.16 using the field transformation equations.

```
In[31]:= R = {x, y, z} ;
```

$$\text{UnitConvert}\left[\left\{0\,\text{T},\ \frac{\gamma\,\beta\,z\,e}{c\,4\,\pi\,\varepsilon_0\,(\text{R.R})^{3/2}},\ -\frac{\gamma\,\beta\,y\,e}{c\,4\,\pi\,\varepsilon_0\,(\text{R.R})^{3/2}}\right\},\ \text{T}\right]$$

Out[31]= $\{0\,\text{T},\ 5.67971\times10^{-7}\,\text{T},\ -3.78647\times10^{-7}\,\text{T}\}$

Mathematica Starter

A.1 CELLS

Mathematica notebooks have two types of cells: "text" cells that can only display what is written and "input" cells that are executable. Input cells are executed by typing (simultaneously)

SHIFT RETURN

which generates the label In[ ]:= with the output going to another input cell with the label Out[ ]= (but it is still an input cell and can be executed).

Example A.1 Calculate 1+1.

```
In[1]:= 1 + 1

Out[1]= 2
```

A semicolon after a line of code means the code will still execute but the output will be suppressed. This is a useful feature for debugging code.

Example A.2 Set $x = 7$ and $y = 2$ and output $x + y$.

```
In[2]:= x = 7; y = 2;
        x + y

Out[3]= 9
```

When a variable is set, its value may be used in other cells until cleared.

A.2 PALETTES

The menu has several extensive palettes that are useful in formatting the input. For example, there is a Writing Assistant that manages cells and fonts. This is useful for quick access to Greek letters. There is a Math Assistant that has templates for operations like division, raising to a power, summation, integration, etc. This makes it easy to enter something like a summation of squares into an input cell in a very clean format.

Example A.3 Sum the squares of integers from 0 to 10.

In[4]:= $\sum_{n=0}^{10} n^2$

Out[4]= 385

A.3 FUNCTIONS

Mathematica functions always begin with a capital letter. When typing in a cell, Mathematica will give autocomplete options for existing functions. Mousing over a function gives extensive documentation for the function's use with examples. The function Clear [ ] clears a variable. It produces no output.

Example A.4 Set $x = 1$ and clear x.

```
In[5]:= x = 1
        Clear[x]
        x
Out[5]= 1

Out[7]= x
```

The function ClearAll[ ] is extremely useful to perform a global clear of everything.

Example A.5 Clear all variables.

```
In[8]:= ClearAll["Global`*"]
```

In[9]:= $\partial_x x^3$

Out[9]= $3\ x^2$

The function Simplify[ ] reduces the result algebraically.

Example A.6 Calculate $\sin^2 x + \cos^2 x$.

```
In[10]:= Simplify[Cos[x]^2 + Sin[x]^2]
Out[10]=
1
```

It can also be written as //Simplify, placed after the code.

Example A.7 Simplify $\frac{x^2-4x+4}{x-2}$.

```
In[11]:= (x^2 - 4 x + 4)/(x - 2) // Simplify
Out[11]=
-2 + x
```

The function D[ ] gives the dervative.

Example A.8 Calculate $\frac{d\cos x}{dx}$.

```
In[12]:= D[Cos[x], x]
Out[12]=
-Sin[x]
```

A user may define a function by placing an underscore after the argument, $f[x\_]$. This allows the function to be evaluated for any value of the argument.

Example A.9 Define the function $f(x) = x^2$ and evaluate it for $x = 2.5$.

```
In[13]:= f[x_] = x^2; f[2.5]
Out[13]=
6.25
```

A.4 RESERVED NAMES

There are a few names that are reserved and may not be user defined. One of them is D which is reserved for differentiation. The reserved names always begin with capital letters. Others include, E , I, and Pi, which stand for the exponential e , imaginary i, and π.

Example A.10 Calculate $e^{i\pi} + 1$.

```
In[14]:= x = E^(I Pi) + 1
Out[14]=
0
```

A double equal sign makes a logical comparison.

Example A.11 Compare Pi π .

```
In[15]:=  Pi == π
Out[15]=
          True
```

A.5 PHYSICAL CONSTANTS AND THEIR UNITS

One can get physical constants in Mathematica by typing simultaneously (in input cell)

CTRL +

and then typing into the natural language box that appears, for example, "speed of light".

Figure A.1 Typing "speed of light" into the natural language box..

Clicking outside the box gives (hopefully) what one was looking for, displayed in standard physics notation.

Figure A.2 Successful procurement of the speed of light.

The physical constant c is stored as a "unit," and it appears in italics with a different shading so you can recognize the difference between a unit and a user-defined variable with the same name. The numerical value is displayed together with units using the function UnitConvert[ ]. The default units will be SI.

Example A.12 Get the numerical value of the speed of light.

```
In[18]:= UnitConvert[ c ]
Out[18]=
        299 792 458 m/s
```

The units to be displayed may be specified. There are two ways to get a unit,
1) typing into the natural language box, and
2) using the function Quantity[ ].

Example A.13 Get the unit miles per second.

```
In[19]:= Quantity["MilesPerSecond"]
Out[19]=
        1 mi/s
```

Example A.14 Get the numerical value of the speed of light in miles per second.

```
In[20]:= x = UnitConvert[c, mi/s]
Out[20]=
```

$$\frac{18\,737\,028\,625}{100\,584}\ \mathrm{mi/s}$$

The function N[ ] will calculate the numerical value to the specified number of significant figures. The value has been stored in the variable x and it remains so until cleared.

Example A.15 Get the numerical value of the previous calculation of the speed of light to three significant figures.

```
In[21]:= N[x, 3]
Out[21]=
```

$$1.86 \times 10^5\ \mathrm{mi/s}$$

Example A.16 Get π to 50 figures.

```
In[22]:= N[π, 50]
Out[22]=
        3.1415926535897932384626433832795028841971693993751
```

The functions NumberForm [ ], and ScientificForm[ ] can also be used to display a decimal answer with specified number of digits.

Other useful physical constants are similarly obtained by typing the following into the natural language box: elementary charge, epsilon_0, planck's constant, hbar, electron mass, proton mass, boltzmann constant, etc. Physical constants and their names are given in App. D.

Mathematica is extremely useful as a calculator because it will automatically check the units of a calculation and report errors.

Example A.17 Try to get the speed of light in kg.

```
In[23]:= UnitConvert[c , kg]

Out[23]=
         $Failed
```

APPENDIX B

Vectors

B.1 NOTATION

Vectors are denoted with curly brackets. One has to be careful in Mathematica with the use of subscripts so not to confuse them with vector components. Subscripts are technically not legal variables and cannot be cleared. The first two lines of code in Ex. B.1 allow the use of subscripts without interference and produce no output.

Example B.1 Define the vector **A** with components (A_x, A_y, A_z).

```
In[1]:= << Notation`
        Symbolize[ParsedBoxWrapper[SubscriptBox["_", "_"]]]
        A = {Ax, Ay, Az}

Out[3]= {Ax, Ay, Az}
```

B.1.1 Scalar Product

The function Dot$[a, b]$ takes the scalar (dot) product of the inserted vectors.

Example B.2 Define the vector **B** with components (B_x, B_y, B_z) and take the dot product $\mathbf{A} \cdot \mathbf{B}$.

```
In[4]:= B = {Bx, By, Bz}; Dot[A, B]

Out[4]= Ax Bx + Ay By + Az Bz
```

The shortcut in standard form is just a period. In ex. B.3 the == does a logical comparison. The conversion to standard form is available from the cell menu.

Example B.3 Show the two ways of writing the dot product are equivalent.

```
In[5]:= A.B == Dot[A, B]

Out[5]= True
```

B.1.2 Vector Product

The function Cross[a,b] takes the vector (cross-) product of the inserted vectors.

Example B.4 Take the product $\mathbf{A} \times \mathbf{B}$.

```
In[6]:= Cross[A, B]
```

Out[6]= $\{-A_z B_y + A_y B_z, A_z B_x - A_x B_z, -A_y B_x + A_x B_y\}$

The shortcut in standard form is a small cross. This and other shortcuts are available in the special characters palette for quick access.

Example B.5 Show the two ways of writing the cross-product are equivalent.

```
In[7]:= A × B == Cross[A, B]

Out[7]= True
```

B.2 DERIVATIVES

B.2.1 Gradient

In Cartesian coordinates, the gradient of a scalar function $f(x,y,z)$ is

$$\nabla f = \frac{\partial f}{\partial x}\,\hat{\mathbf{x}} + \frac{\partial f}{\partial y}\,\hat{\mathbf{y}} + \frac{\partial f}{\partial z}\,\hat{\mathbf{z}}.$$

The gradient is obtained with the function Grad[f,{x,y,z}] which differentiates f w.r.t. (x,y,z). In Ex. B.6 the gradient function has been put into standard form.

Example B.6 Calculate $\nabla(\sqrt{x^2+y^2+z^2})$.

```
In[8]:= ∇{x,y,z} (√(x² + y² + z²))

Out[8]= {x/√(x² + y² + z²), y/√(x² + y² + z²), z/√(x² + y² + z²)}
```

B.2.2 Divergence

In Cartesian coordinates, the divergence of a vector function $\mathbf{A}(x,y,z)$ is

$$\nabla \cdot \mathbf{A} = \frac{\partial A_x}{\partial x} = \frac{\partial A_y}{\partial y} = \frac{\partial A_z}{\partial z}.$$

The divergence is obtained with the function Div[f,{x,y,z}], which differentiates f w.r.t. (x,y,z). In Ex. B.7 the divergence function has been put into standard form.

Example B.7 Calculate $\nabla \cdot (x,y,z)$.

```
In[9]:= r = {x, y, z}; ∇{x,y,z} · r

Out[9]= 3
```

B.2.3 Curl

In Cartesian coordinates, the curl of a vector function $\mathbf{A}(x,y,z)$ is

$$\nabla \times \mathbf{A} = \left(\frac{\partial A_z}{\partial y} - \frac{\partial A_y}{\partial z}\right)\hat{\mathbf{x}} + \left(\frac{\partial A_x}{\partial z} - \frac{\partial A_z}{\partial x}\right)\hat{\mathbf{y}} + \left(\frac{\partial A_y}{\partial x} - \frac{\partial A_x}{\partial y}\right)\hat{\mathbf{z}}.$$

The curl is obtained with the function Curl[f,{x,y,z}] which differentiates f w.r.t. (x,y,z). In Ex. B.8 the curl function has been put into standard form.

Example B.8 Calculate $\nabla \times (-y,x,0)$.

```
In[10]:= f = {-y, x, 0}; ∇{x,y,z} × f

Out[10]= {0, 0, 2}
```

B.3 LINE INTEGRAL

The line integral of a vector function **f** is written $\int_{\mathbf{a}}^{\mathbf{b}} d\boldsymbol{\ell}\cdot\mathbf{f}$ where **a** and **b** are vector positions. In general, the integral depends on the path. Consider the line integral of a vector function,

$$\mathbf{f} = (y^2, 2xy - 2x, 0),$$

along two different paths. Path 1 goes in a straight line ($y = x$) from $\mathbf{a} = (1,1,0)$ to $\mathbf{b} = (4,4,0)$, and path 2 goes along $y = 1$ and then $x = 3$ (Fig. B.1).

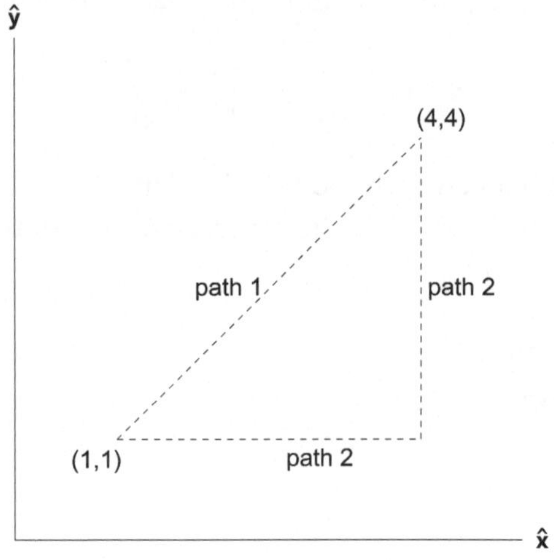

Figure B.1 The line integral is taken along two different paths.

Along path 1, $y = x$ and $dy = dx$. The vector components are obtained by the dot product with the unit vectors, $f_x = \mathbf{f}\cdot\hat{\mathbf{x}}$ and $f_y = \mathbf{f}\cdot\hat{\mathbf{y}}$.

Example B.9 Calculate $\int_{\mathbf{a}}^{\mathbf{b}} d\boldsymbol{\ell}\cdot\mathbf{f}$ along path 1.

```
In[11]:= f = {y², 2 x (y - 1), 0};
         ∫₁⁴ (f.{1, 0, 0} /. y → x) dx + ∫₁⁴ (f.{0, 1, 0} /. y → x) dx

Out[12]= 48
```

Along path 2, $y = 1$ for the horizontal part and $x = 4$ for the vertical part.

Example B.10 Calculate $\int_{\mathbf{a}}^{\mathbf{b}} d\boldsymbol{\ell}\cdot\mathbf{f}$ along path 2.

```
In[13]:= ∫_1^4 (f.{1, 0, 0} /. y → 1) dx + ∫_1^4 (f.{0, 1, 0} /. x → 4) dy
Out[13]= 39
```

There is a fundamental condition that makes the line integral path-independent. This condition is that the curl of the vector function be zero. Consider the function

$$\mathbf{f} = \frac{x}{(x^2+y^2+z^2)^{3/2}}\,\hat{\mathbf{x}} + \frac{y}{(x^2+y^2+z^2)^{3/2}}\,\hat{\mathbf{y}},$$

which is the coordinate dependence of the electric field of a point charge.

Example B.11 Calculate the curl of **f**.

```
In[14]:= f = {x/(x^2 + y^2)^(3/2), y/(x^2 + y^2)^(3/2), 0};
         ∇{x,y,z} × f
Out[15]= {0, 0, 0}
```

Example B.12 Calculate $\int_{\mathbf{a}}^{\mathbf{b}} d\boldsymbol{\ell}\cdot\mathbf{f}$ along path 1.

```
In[16]:= ∫_1^4 (f.{1, 0, 0} /. y → x) dx + ∫_1^4 (f.{0, 1, 0} /. y → x) dx
Out[16]= 3/(4 √2)
```

Example B.13 Calculate $\int_{\mathbf{a}}^{\mathbf{b}} d\boldsymbol{\ell}\cdot\mathbf{f}$ along path 2.

```
In[17]:= ∫_1^4 (f.{1, 0, 0} /. y → 1) dx + ∫_1^4 (f.{0, 1, 0} /. x → 4) dy
Out[17]= 3/(4 √2)
```

B.4 FLUX INTEGRAL

An area integral $\int d\mathbf{a} \cdot \mathbf{f}$ is often referred to as a "flux." The direction of the differential vector $d\mathbf{a}$ is taken to be perpendicular to the surface,

$$d\mathbf{a} = da\,\hat{\mathbf{n}},$$

where $\hat{\mathbf{n}}$ is the unit vector normal to the surface. There are two possible directions for $\hat{\mathbf{n}}$. If the surface is closed, the direction is taken to be outward. If the surface is not closed, the direction must be specified.

Consider the function

$$\mathbf{f} = (2x^2 z, x + 7, yz^2 - 3y)$$

integrated over a square in the $x-y$ plane at $z = 2$ as indicated in Fig. B.2.

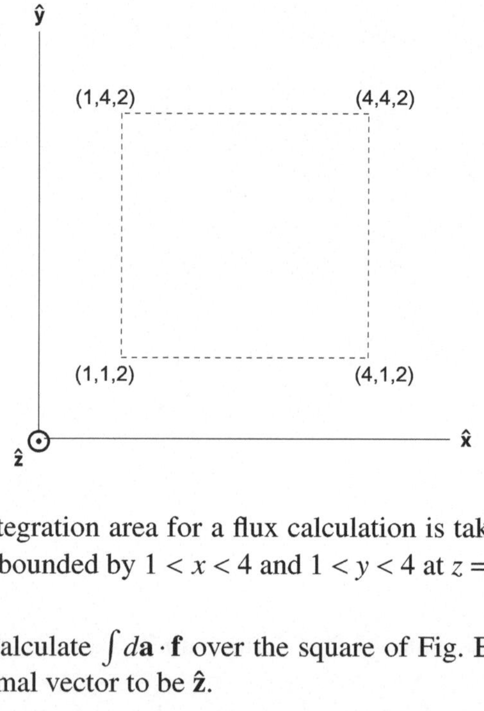

Figure B.2 The integration area for a flux calculation is taken to be a square in the $x-y$ plane bounded by $1 < x < 4$ and $1 < y < 4$ at $z = 2$.

Example B.14 Calculate $\int d\mathbf{a} \cdot \mathbf{f}$ over the square of Fig. B.2, taking the direction of the normal vector to be $\hat{\mathbf{z}}$.

```
In[18]:= f = {2 x^2 z, x + 7, y (z^2 - 3)};
         Integrate[Integrate[f.{0, 0, 1}, {x, 1, 4}],
           {y, 1, 4}] /. z → 2

Out[19]= 45/2
```

B.5 DIVERGENCE THEOREM

The divergence theorem states that

$$\int dv\, \nabla \cdot \mathbf{E} = \oint d\mathbf{a} \cdot \mathbf{E},$$

where the integral on the right is over the closed surface that surrounds the volume that is integrated on the left. It holds for any function and can be proven mathematically.

Example B.15 Calculate the volume integral over a unit cube extending from $-1 < x < 1$, $-1 < y < 1$, and $-1 < z < 1$ of the divergence of **f**.

```
In[20]:= ClearAll["Global`*"] ;
         f[x_, y_, z_] = {x^3 y^2, x y^3, Sin[z]};
         ∫_{-1}^{1} (∫_{-1}^{1} (∫_{-1}^{1} ∇_{x,y,z} · f[x, y, z] dx) dy) dz

Out[21]= 8/3 + 8 Sin[1]
```

Example B.16 Calculate the the flux of **f** on the unit cube.

```
In[22]:= f1 = ∫_{-1}^{1} (∫_{-1}^{1} f[x, y, z].{-1, 0, 0} dy) dz /. x → -1;
         f2 = ∫_{-1}^{1} (∫_{-1}^{1} f[x, y, z].{1, 0, 0} dy) dz /. x → 1;
         f3 = ∫_{-1}^{1} (∫_{-1}^{1} f[x, y, z].{0, -1, 0} dx) dz /. y → -1;
         f4 = ∫_{-1}^{1} (∫_{-1}^{1} f[x, y, z].{0, 1, 0} dx) dz /. y → 1;
         f5 = ∫_{-1}^{1} (∫_{-1}^{1} f[x, y, z].{0, 0, -1} dx) dy /. z → -1;
         f6 = ∫_{-1}^{1} (∫_{-1}^{1} f[x, y, z].{0, 0, 1} dx) dy /. z → 1;
         f1 + f2 + f3 + f4 + f5 + f6

Out[27]= 8/3 + 8 Sin[1]
```

B.6 STOKES' THEOREM

Stokes's theorem states that if one takes the curl of any vector function **f** and then calculates the flux of that curl through some area, then the result is equal to the line integral around the boundary that encloses the area,

$$\int d\mathbf{a} \cdot (\nabla \times \mathbf{f}) = \oint d\boldsymbol{\ell} \cdot \mathbf{E}.$$

Consider the function

$$\mathbf{f} = x^2 y\, \hat{\mathbf{x}} + zy^2\, \hat{\mathbf{y}} + xy\, \hat{\mathbf{z}}.$$

Example B.17 Calculate the flux of $\nabla \times \mathbf{f}$ through a square in the $x-y$ plane bounded by $0 < x < 1$ and $0 < y < 1$.

```
In[28]:= ClearAll["Global`*"];
         f[x_, y_, z_] = {x^2 y, z y^2, x y};
         ∫_0^1 (∫_0^1 ∇{x,y,z} × f[x, y, z].{0, 0, 1} dx) dy

Out[29]= -1/3
```

Example B.18 Calculate the line integral of **f** around the perimeter of the square.

```
In[30]:= z = 0;
         ∫_0^1 (f[x, y, z].{1, 0, 0} /. y → 0) dx +
          ∫_0^1 (f[x, y, z].{0, 1, 0} /. x → 1) dy +
          ∫_1^0 (f[x, y, z].{1, 0, 0} /. y → 1) dx +
          ∫_1^0 (f[x, y, z].{0, 1, 0} /. x → 0) dy

Out[30]= -1/3
```

APPENDIX C

Spherical and Cylindrical Coordinates

C.1 SPHERICAL COORDINATES

Spherical coordinates are described with unit vectors $\hat{\mathbf{r}}$, $\hat{\boldsymbol{\theta}}$, $\hat{\boldsymbol{\phi}}$ that are not constant. The variable r is the distance to the origin in an arbitrary direction and has a range $0 \le r \le \infty$. The polar angle θ measured from the z-axis has a range $0 \le \theta \le \pi$. The azimuthal angle ϕ measured in the $x-y$ plane has a range $0 \le \phi \le 2\pi$. The unit vectors are orthogonal (see Figs. C.1 and C.2) and satisfy

$$\hat{\mathbf{r}} \times \hat{\boldsymbol{\theta}} = \hat{\boldsymbol{\phi}},$$

$$\hat{\boldsymbol{\phi}} \times \hat{\mathbf{r}} = \hat{\boldsymbol{\theta}},$$

and

$$\hat{\boldsymbol{\theta}} \times \hat{\boldsymbol{\phi}} = \hat{\mathbf{r}},$$

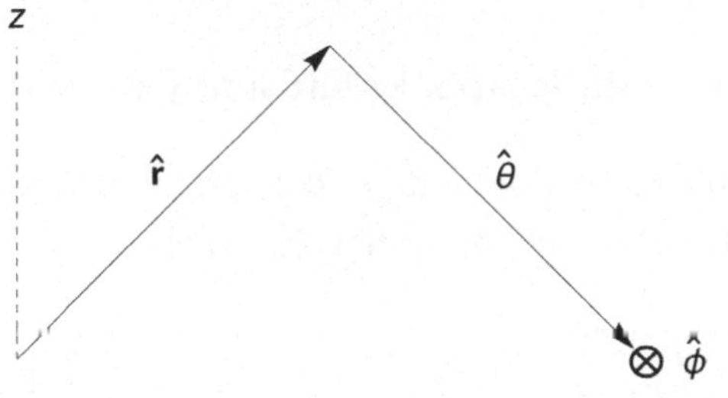

Figure C.1 Spherical-coordinate unit vectors are shown in the $r-z$ plane. The $r-z$ plane depends on the direction of $\hat{\mathbf{r}}$.

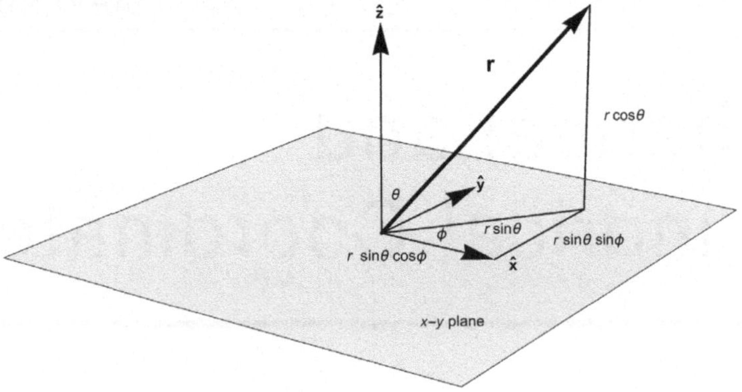

Figure C.2 In spherical coordinates (r,θ,ϕ), the polar angle (θ) is the angle between the vector $\mathbf{r}$ and the z-axis and the azimuthal angle (ϕ) is the angle in the $x-y$ plane.

While spherical coordinates are convenient to describe a geometry with spherical symmetry, one must be very mindful that the unit vectors are not constant. For example, this makes derivatives non-trivial to calculate. Most of the time it is easiest to work in Cartesian coordinates (unit vectors $\hat{\mathbf{x}}$, $\hat{\mathbf{y}}$, $\hat{\mathbf{z}}$) using the spherical variables (r,θ,ϕ). These are obtained with FromSphericalCoordinates[{r,θ,ϕ}].

Example C.1 Get Cartesian coordinates (x,y,x) in terms of spherical variables r,θ,ϕ.

```
In[1]:= FromSphericalCoordinates[{r, θ, ϕ}]

Out[1]= {r Cos[ϕ] Sin[θ], r Sin[θ] Sin[ϕ], r Cos[θ]}
```

Example C.1 says that

$$r\,\hat{\mathbf{r}} = r\sin\theta\cos\phi\,\hat{\mathbf{x}} + r\sin\theta\sin\phi\,\hat{\mathbf{y}} + r\cos\theta\,\hat{\mathbf{z}}.$$

To go in the other direction and get the spherical variables from (x,y,x), for example, use ToSphericalCoordinates[{x,y,x}].

Example C.2 Get spherical coordinates (r,θ,ϕ) in terms of Cartesian variables x,y,z.

```
In[2]:= ToSphericalCoordinates[{x, y, z}]

Out[2]= {√(x² + y² + z²), ArcTan[z, √(x² + y²)], ArcTan[x, y]}
```

Example C.2 says that

$$x\,\hat{\mathbf{x}}+y\,\hat{\mathbf{y}}+z\,\hat{\mathbf{z}}=\sqrt{x^2+y^2+z^2}\,\hat{\mathbf{r}}+\tan^{-1}\left(\frac{\sqrt{x^2+y^2}}{z}\right)\hat{\boldsymbol{\theta}}+\tan^{-1}\left(\frac{y}{x}\right)\hat{\boldsymbol{\phi}}$$

To get a Cartesian unit vectors in terms of spherical unit vectors, use TransformedField["Cartesian"→ "Spherical",f,$\{x,y,x\} \rightarrow \{r,\theta,\phi\}$], where f is the vector field to be transformed. Since the calculation is repetitive, several shortcuts can be used. The symbol # is an abbreviation for the function Slot which makes a substitution for the function after the &, and /@ is an abbreviation for the function Map. The function IdentityMatrix[n] is a matrix of unit vectors for n dimensions. The function Column outputs the vectors in a column.

Example C.3 Get Cartesian unit vectors $\hat{\mathbf{x}},\hat{\mathbf{y}},\hat{\mathbf{z}}$ in terms of spherical unit vectors $\hat{\mathbf{r}},\hat{\boldsymbol{\theta}},\hat{\boldsymbol{\phi}}$ and angles.

```
In[3]:= TransformedField["Cartesian" → "Spherical", #,
            {x, y, z} → {r, θ, ϕ}] & /@ IdentityMatrix[3] // Column

        {Cos[ϕ] Sin[θ], Cos[θ] Cos[ϕ], -Sin[ϕ]}
Out[3]= {Sin[θ] Sin[ϕ], Cos[θ] Sin[ϕ], Cos[ϕ]}
        {Cos[θ], -Sin[θ], 0}
```

Example C.3 says that

$$\hat{\mathbf{x}}=\sin\theta\cos\phi\,\hat{\mathbf{r}}+\cos\theta\cos\phi\,\hat{\boldsymbol{\theta}}-\sin\phi\,\hat{\boldsymbol{\phi}},$$

$$\hat{\mathbf{y}}=\sin\theta\sin\phi\,\hat{\mathbf{r}}+\cos\theta\sin\phi\,\hat{\boldsymbol{\theta}}+\cos\phi\,\hat{\boldsymbol{\phi}},$$

and

$$\hat{\mathbf{z}}=\cos\theta\,\hat{\mathbf{r}}-\sin\theta\,\hat{\boldsymbol{\theta}}.$$

Example C.4 Get spherical unit vectors $\hat{\mathbf{r}},\hat{\boldsymbol{\theta}},\hat{\boldsymbol{\phi}}$ in terms of Cartesian unit vectors $\hat{\mathbf{x}},\hat{\mathbf{y}},\hat{\mathbf{z}}$ and angles.

```
In[4]:= Simplify[
          TransformedField["Cartesian" → "Spherical", #,
             {x, y, z} → {r, θ, ϕ}] & /@
           TransformedField["Spherical" → "Cartesian", #,
             {r, θ, ϕ} → {x, y, z}], {r > 0, 0 < θ < π}] & /@
        IdentityMatrix[3] // Column

        {Cos[ϕ] Sin[θ], Sin[θ] Sin[ϕ], Cos[θ]}
Out[4]= {Cos[θ] Cos[ϕ], Cos[θ] Sin[ϕ], -Sin[θ]}
        {-Sin[ϕ], Cos[ϕ], 0}
```

Example C.4 says that

$$\hat{\mathbf{r}} = \sin\theta\cos\phi\,\hat{\mathbf{x}} + \sin\theta\sin\phi\,\hat{\mathbf{y}} + \cos\theta\,\hat{\mathbf{z}},$$

$$\hat{\boldsymbol{\theta}} = \cos\theta\cos\phi\,\hat{\mathbf{x}} + \cos\theta\sin\phi\,\hat{\mathbf{y}} - \sin\theta\,\hat{\mathbf{z}},$$

$$\hat{\boldsymbol{\phi}} = -\sin\phi\,\hat{\mathbf{x}} + \cos\phi\,\hat{\mathbf{y}}.$$

C.2 CYLINDRICAL COORDINATES

Cylindrical coordinates keep the z-axis fixed, using unit vectors $\hat{\mathbf{r}}$, $\hat{\boldsymbol{\phi}}$, $\hat{\mathbf{z}}$. The variable r is the distance to the z-axis and has a range $0 \le r \le \infty$. The azimuthal angle ϕ measured in the $x-y$ plane has a range $0 \le \phi \le 2\pi$. The unit vectors are orthogonal (see Figs. C.3 and C.4) satisfy

$$\hat{\mathbf{r}} \times \hat{\boldsymbol{\phi}} = \hat{\mathbf{z}},$$

$$\hat{\boldsymbol{\phi}} \times \hat{\mathbf{z}} = \hat{\mathbf{r}},$$

and

$$\hat{\mathbf{z}} \times \hat{\mathbf{r}} = \hat{\boldsymbol{\phi}},$$

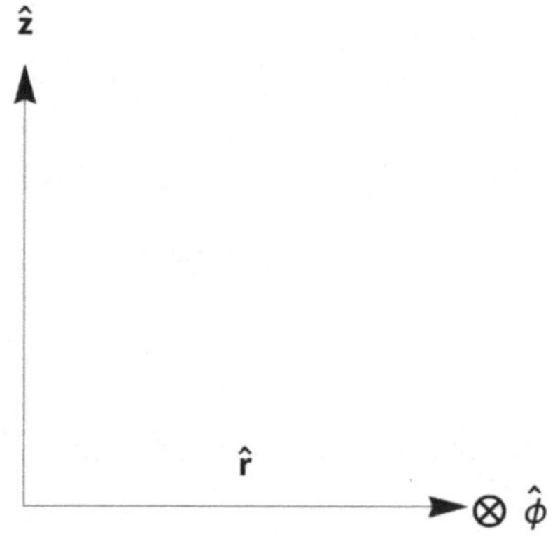

Figure C.3 Cylindrical-coordinate unit vectors are shown in the $r-z$ plane. The $r-z$ plane depends on the direction of $\hat{\mathbf{r}}$.

Example C.5 Get Cartesian coordinates (x, y, x) in terms of cylindrical variables r, ϕ, z.

```
In[5]:= CoordinateTransform["Cylindrical" → "Cartesian", {r, ϕ, z}]

Out[5]= {r Cos[ϕ], r Sin[ϕ], z}
```

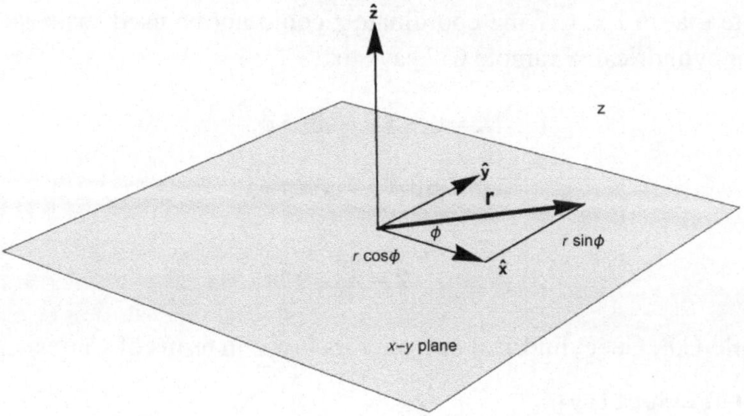

Figure C.4 In cylindrical coordinates (r,ϕ,z), the azimuthal angle (ϕ) is the angle in the $x-y$ plane.

Example C.5 says that

$$r\,\hat{\mathbf{r}} + z\,\hat{\mathbf{z}} = r\sin\phi\,\hat{\mathbf{x}} + r\sin\phi\,\hat{\mathbf{y}} + z\,\hat{\mathbf{z}}.$$

Example C.6 Get cylindrical coordinates (r,ϕ,z) in terms of Cartesian variables x,y,z.

```
In[6]:= CoordinateTransform["Cartesian" → "Cylindrical", {x, y, z}]

Out[6]= {√(x² + y²), ArcTan[x, y], z}
```

Example C.6 says that

$$x\,\hat{\mathbf{x}} + y\,\hat{\mathbf{y}} + z\,\hat{\mathbf{z}} = \sqrt{x^2+y^2}\,\hat{\mathbf{r}} + \tan^{-1}\left(\frac{y}{x}\right)\hat{\boldsymbol{\phi}} + z\,\hat{\mathbf{z}}$$

Example C.7 Get Cartesian unit vectors $\hat{\mathbf{x}}, \hat{\mathbf{y}}, \hat{\mathbf{z}}$ in terms of cylindrical $\hat{\mathbf{r}}, \hat{\boldsymbol{\phi}}, \mathbf{z}$.

```
In[7]:= TransformedField["Cartesian" → "Cylindrical", #,
            {x, y, z} → {r, ϕ, z′}] & /@ IdentityMatrix[3] // Column

        {Cos[ϕ], -Sin[ϕ], 0}
Out[7]= {Sin[ϕ], Cos[ϕ], 0}
        {0, 0, 1}
```

Note that in Ex. C.7 the coordinate z could not be used twice so z' was used for cylindrical. Example C.7 says that

$$\hat{\mathbf{x}} = \cos\phi\,\hat{\mathbf{r}} - \sin\phi\,\hat{\boldsymbol{\phi}},$$

$$\hat{\mathbf{y}} = \sin\phi\,\hat{\mathbf{r}} + \cos\phi\,\hat{\boldsymbol{\phi}},$$

and

$$\hat{\mathbf{z}} = \hat{\mathbf{z}}'.$$

Example C.8 Get cylindrical unit vectors $\hat{\mathbf{r}}, \hat{\boldsymbol{\phi}}, \mathbf{z}$ in terms of Cartesian $\hat{\mathbf{x}}, \hat{\mathbf{y}}, \hat{\mathbf{z}}$.

```
In[8]:= FullSimplify[
          TransformedField["Cartesian" → "Cylindrical", #,
            {x, y, z} → {r, ϕ, z′}] & /@
           TransformedField["Cylindrical" → "Cartesian", #,
            {r, ϕ, z′} → {x, y, z}], {r > 0, 0 < θ < π}] & /@
        IdentityMatrix[3] // Column

        {Cos[ϕ], Sin[ϕ], 0}
Out[8]= {-Sin[ϕ], Cos[ϕ], 0}
        {0, 0, 1}
```

Example C.8 says that

$$\hat{\mathbf{r}} = \cos\phi\,\hat{\mathbf{x}} + \sin\phi\,\hat{\mathbf{y}},$$

$$\hat{\boldsymbol{\phi}} = -\sin\phi\,\hat{\mathbf{x}} + \cos\phi\,\hat{\mathbf{y}},$$

$$\hat{\mathbf{z}}' = \hat{\mathbf{z}}.$$

APPENDIX D

Physical Constants

Physical constants may be called in two ways. One way is to use the function Quantity[ ] with the argument equal to the name of the constant. A second way which makes a much cleaner look to the code is to use the defined symbol obtained from the natural language box as described in A.5. Executing Quantity[ ] produces the an output identical to that of the natural language box.

Example D.1 Compare the elementary charge as obtained from the Quantity[ ] function and the natural language box.

```
In[1]:= Quantity["ElementaryCharge"] == e

Out[1]= True
```

The names of the fundamental constants used in this book with their symbols and values are shown in D.1 and derived combinations are shown in D.2.

Particle masses may be acquired with either the function Quantity[ ] or the natural language box as described in A.5.

Example D.2 Get the numerical value of the electron mass using the natural language box.

```
In[2]:= ScientificForm[UnitConvert[ electron PARTICLE [ mass ]], 4]

Out[2]//ScientificForm=
        9.109×10^-31 kg
```

Particle masses are displayed in D.3.

Table D.1 Mathematica names (symbol) and numerical values for physical constants.

Physical Constant	Value
ElementaryCharge (e)	1.60218×10^{-19} C
ElectricConstant (ε_0)	$8.85418781 \times 10^{-12}$ C/(m V)
MagneticConstant (μ_0)	$1.256637062 \times 10^{-6}$ m T/A
SpeedOfLight (c)	299 792 458 m/s
PlanckConstant (h)	4.13567×10^{-15} s eV
GravitationalConstant (G)	6.6743×10^{-11} $m^3/(kg\ s^2)$
BoltzmannConstant (k)	0.0000861733 eV/K
AvogadroNumber (N_0)	6.02214×10^{23}

Table D.2 Mathematica names (symbol) and numerical values for derived constants.

Derived Constant	Value
ReducedPlanckConstant ($\hbar$)	6.58212×10^{-16} s eV
FineStructureConstant (α)	0.007297352569
ElectronComptonWavelength(λ_e)	$2.42631024 \times 10^{-12}$ m
BohrRadius (a_0)	0.0529177211 nm
RydbergConstant (R_∞)	1.09737×10^{7} per meter
BohrMagneton (μ_B)	0.0000578838 eV/T
StefanBoltzmannConstant (σ)	5.67037×10^{-8} $W/(m^2 K^4)$

Table D.3 Mathematica names (symbol) and numerical values for particle masses.

Particle	Mass
ElectronMass (m_e)	0.510999 MeV/c^2
MuonMass (m_μ)	105.658 MeV/c^2
ProtonMass (m_p)	938.272 MeV/c^2
NeutronMass (m_n)	939.565 MeV/c^2
DeuteronMass (m_d)	1875.61 MeV/c^2
AlphaParticleMass (m_α)	3727.38 MeV/c^2

Table D.4 Common names (symbol) and numerical values for sky objects.

Object	Mass
mass of Moon (Moon [mass])	7.3459×10^{22} kg
mass of Earth (Earth [mass])	5.9722×10^{24} kg
solar mass (Sun [mass])	1.98844×10^{30} kg
neutron star mass ((1.1 to 2.1) $M_\odot$)	$(2.18724 \times 10^{30}$ to $4.17564 \times 10^{30})$ kg
Milky Way mass (Milky Way [mass])	3.06214×10^{42} kg
mass of universe (m_U])	1.51184×10^{53} kg

Index

$Assumptions, Mathematica, 9

alternating current, 122
Ampère's law, steady current, 89
Ampere's law, steady current, 79
auxiliary field **H**, 146
average electric field, 25

ball of charge, 23, 42, 52
Biot-Savart law, 63, 66, 83
bound charge density, 132
bound current, 141
bound surface charge, 132
bound surface current, 144
boundary condition, 149, 161
Brewster angle, 167

capacitance, 57
capacitor, charging, 155
capacitor, moving, 180
capacitor, stored energy, 61
cells, Mathematica, 191
circuit, LR, 122
circuit, LRC, 127
circuit, RC, 121, 125
circuit, resistor, 117, 118
Clear, Mathematica, 192
coax cable, 111
complex wave notation, 159
conductivity, 113
conductors, 56
continuity equation, 67, 151, 156
coordinate system, 4
coulomb unit, 2
Coulomb's law, 1, 4, 6
Cross, Mathematica, 198
curl of **A**, 91, 93, 94
curl of **B**, 89
curl of **E**, 26, 45
curl of cross product, 83
curl of curl, 157
curl of scalar times vector, 95
Curl, Mathematica, 199
current, circular, 71, 82
current, cylindrical, 70, 86
current, line, 68, 85, 94
current, sheet, 87, 93
current, square, 80
cylindrical coordinates, 208

Delta-Y transform, 120
derivative, Mathematica, 193
dielectric, 136
dielectric ball, 137
dielectric boundary, 139
dielectric, electric field in, 137
dielectric, inside capacitor, 138
dipole field, electric, 135
dipole potential, electric, 131
dipole, electric, 53
dipole, magnetic, 75, 100
disk of charge, 19, 50
displacement vector, **D**, 136
Div, Mathematica, 199
divergence, 84, 199
divergence of **B**, 65
divergence of **E**, 38, 39
divergence of cross product, 153
divergence of curl, 91, 151

divergence theorem, 84, 203
Dot, Mathematica, 197
drift speed, 114

electric charge, 1
electric constant, ε_0, 1, 3
electric field, 12
electric potential, 45, 48
electromagnetic wave, 151
elementary charge, e, 1, 2
EMF, $\mathcal{E}$, 103
energy per volume, 153
energy, magnetic, 110

Faraday's law, 103
Fermi speed, 114
field transformation, 186
fields in materials, 131
flux integral, 202
flux rule, 103
flux, electric, 29, 32
flux, magnetic, 65, 105
four-vector, energy-momentum, 176
four-vectors, 172
Fresnel equations, 166
FullSimplify, Mathematica, 16

Gauss's law, 29, 39
Grad, Mathematica, 47, 198
gradient, 47, 83

inductance, mutual, 105
inductance, self, 109
inverse square law, 2

kinetic energy, 178
Kirchhoff's rules, 117

length contraction, 171, 175
Lenz's law, 103
line integral, 200
line integral of **B**, 79, 82
line of charge, 13, 40, 48
Lorentz force, 64, 152
Lorentz transformation, 173

magnetic charge, 152
magnetic constant, μ_0, 1, 64
magnetic field, 63
magnetic force, 63
magnetization vector, **M**, 140
magnetized ball, 143, 148
Maxwell equation, 38, 65, 89, 104, 151
mean free path, 115
metric tensor, g, 172
moving point charge, 184, 189

Newton's 3rd law, 4
notation, 4
numerical integration, 72

Ohm's law, 113

palettes, Mathematica, 192
parallel plates, 58
permittivity, relative, 136
physical constants, 211
physical constants, Mathematica, 194
plane of charge, 41
plane wave, 157, 158
point charge, 12, 46
polarization vector, **P**, 131
polarized ball, 134
polarized cube, 132
polarized wave, 165
potential difference, 45
Poynting vector, 152

reflection, 161
reflection coefficient, 163

refraction, 161
reserved names, Mathematica, 193
resistivity, 113
resistors, 116
ring of charge, 17, 50

scalar product, 197
semicolon, Mathematica, 3, 191
Simplify, Mathematica, 16
Snell's law, 165
solenoid, 73, 87, 92, 108
special relativity, 171
speed of light, 67
sphere of charge, 21, 51, 97
spherical coordinates, 205
steady current, 181
Stokes' theorem, 46, 80, 92, 105, 204
stored energy, 60
stored magnetic energy, 154
strength of electric force, 27
strength of gravity, 27
substitute, Mathematica, 6
superposition principle, 6
susceptibility, electric, 136
susceptibility, magnetic, 147

tesla unit, 91
time average, 159
time dilation, 171, 174
toroid, 88
total energy, 177
transmission coefficient, 163

UnitConvert, Mathematica, 2
units, Mathematica, 195

vector Laplacian, 157
vector potential, **A**, 91
vector product, 198
vector, Mathematica, 5, 197
velocity addition, 179

wave equation, 156
waves in matter, 159
wire segment, 96